THE POCKET GUIDE TO
MUSHROOMS
AND
OTHER FUNGI

THE POCKET GUIDE TO
MUSHROOMS
AND
OTHER FUNGI

GEOFFREY KIBBY

**DRAGON'S
WORLD**

Dragon's World Ltd
Limpsfield
Surrey RH8 0DY
Great Britain

First published by Dragon's World Ltd 1991

Editor: Diana Steedman
Designer: Mel Raymond
Editorial Director: Pippa Rubinstein
Art Director: Dave Allen

The catalogue record for this book is available from the British
Library

ISBN 1 85028 121 1

Printed in Singapore

Contents

Introduction

In Britain there is a tradition, both professional and amateur, of studying fungi (known as mycology) which has provided a legacy of scientific texts and field guides describing the mushrooms of our woods and fields. As a result, we know more about the mushrooms and other fungi found in Britain than those found anywhere else in the world. Yet despite this intensive study – covering more than two centuries – we are still discovering new facts and new species every year.

Estimates of the number of different species of larger fungi found in Britain vary, but there are at least 2000. Most of them are very specific in their choice of habitat and requirements for growth and reproduction. The range of habitats vary from the extremes of bleak, highland mountain tops to the salty slacks behind seashore sand-dunes; from the humid, mossy slopes of deciduous forests to city parks and gardens. There is nowhere that mushrooms cannot be searched for and it is for exactly this reason that the amateur mycologist can contribute so much to our knowledge of these fascinating organisms.

By studying the more unusual or neglected sites amateurs can contribute information on distribution, and discover species new to Britain or even new to science. Any addition to the incomplete data on where some can be found will be valuable. Whilst this pocket field guide cannot show all the species known to occur, it does illustrate and describe over 420 of the most important, distinctive or easily identifiable species together with those that are important edibles or are poisonous. Some rare species are included because any day's ramble can be enlivened by the hope of finding some little-known fungi.

What is a Mushroom?

Mushrooms are any of the larger fungi that you may find on a walk through the woods, fields or even in your garden. The words mushroom and toadstool, both commonly used, mean the same thing and give no guide to their edibility or otherwise. They have no scientific meaning and are simply names used over several hundred years, in many ways, by different people, in different countries. The larger fungi (anything over about 0.6cm) are the fruiting bodies or reproductive bodies formed to produce and distribute their spores. The bulk of the fungus consists of almost invisible threads called mycelium, which run through the soil, wood or other substrate on which it grows and feeds.

The spores form in two ways. In most of the fungi shown in this book – the Basidiomycetes – they form on the outer surface of club-like cells called basidia. In the cup-fungi and related forms – the Ascomycetes – they form inside long cells called asci and fire out like bullets.

A glance at the illustrations will show that mushrooms come in a bewildering variety of shapes and colours, and that some method must be adopted to increase your chance of identifying your collections.

How to Use this Book

First you should become familiar with the different parts of the mushrooms you may find, and with the technical terms used to describe those parts. The pictorial keys show typical mushroom types and their various parts – in particular the area where the spores form, the **hymenium.** This may be on flaps called gills or lamellae; lining the interior of tiny tubes; on spines or teeth; or on an irregular, smooth to wrinkled surface. If your mushroom has gills it is very important to note how those gills join to the stem. They may be completely **free** of the stem; **adnate,** when they join the stem for part or all their width; **decurrent,** where the gills meet the stem and run down it for some distance; or **sinuate,** where the gill curves up as it touches the stem.

Make accurate notes on the other features of your mushroom. Is there a ring present on the stem? Does it have a sac or volva at the base? Does the mushroom change colour if you bruise it or cut it open? Is there a distinctive odour or taste present? To taste a mushroom (which is very important in some groups) put only the smallest piece on the tip of the tongue and roll it around, never swallow it. If done carefully there is no risk, although it is usual to learn the poisonous species by sight to avoid unnecessary tasting of these.

The most important character is the colour of the spores in a deposit and to obtain this you must make a spore-print. This is easy to do: simply place a mushroom cap with the gills or tubes or other spore-producing surface down on a sheet of white paper and cover with a cup or other container for a few hours. After this period carefully lift the mushroom cap and underneath you should have a good deposit of spores.

The deposit should be left to dry for a few minutes and then scraped together to observe the colour. This will fall into one of four broad categories. The largest group is the white to pale colours, including pale creams, yellows and ochres to lilac or even greenish. Then there are the brown spores that vary from bright rust-brown through ochre-brown and a dull, earthy brown, often called cigar-brown. The third group is the black, blackish-brown, purplish-brown to purple group, and finally there are the pink, or deep salmon-coloured spores.

The Descriptions

Once you think you have found an illustration that closely matches your specimen, then read the description carefully. All the features should match your specimen – if not, then remember there are many more species than can be shown here, and you may have a closely related species. Is your specimen young and in good condition? Old or poorly collected material is difficult to identify.

Microscopic details of the spores included for each species give the range of the length and breadth, i.e. spores 7–8 x 4–5μm, which means the spores vary from 7 to 8 micrometers in length by 4 to 5 micrometers in width. A micrometer (often called a micron) is one thousandth of a millimetre.

You do not have to have a microscope to enjoy collecting and identifying mushrooms, but access to one will greatly increase the number of species you can name, which is a fascinating hobby in itself. Some basic chemical reactions are given in some descriptions and if you can obtain some of these chemicals then they can be very helpful. Everyday chemicals are available in most households. For example, ammonia is present in many household glass cleaners, and these are adequate for testing. Others are more specialized and their formulas are given in the Appendix.

Collecting

Only collect what you need to study, never over-collect and try not to disturb the habitat. The fungus mycelium will remain undisturbed and so picking the mushroom will not endanger the fungus. Always collect all the mushroom, especially the base of the stem, which is very important in some poisonous groups. Wrap the specimen in a sheet of waxed paper, rolling it carefully and twisting the ends like a sweet: this will stop the fungus from being crushed and will retain moisture. Never collect in a plastic bag! Mushrooms soon sweat and become a soggy mess in such bags.

Join your local mushroom group and go on their forays where you can learn from more experienced collectors and experts. Apart from your local natural history society, the British Mycological Society caters for both the amateur and professional and organizes meetings and forays all over the country.

Edibility

Because the eating of wild mushrooms is becoming

increasingly popular, the edibility of each species wherever known or as suggested in other specialist literature is indicated. However, it must be stressed that individual reactions to any food can vary, even to well-known edibles, and the author and publishers cannot assume responsibility for the consequences of readers eating wild mushrooms. Do not eat any mushroom without first getting expert opinion on the identification.

Some mushrooms are dangerously poisonous or even deadly, and this is indicated by a skull and crossbones symbol beside the illustration. Many people are poisoned each year; do not take unnecessary risks. Join a mushrooming group, or ask an expert, if you wish to learn about eating mushrooms.

Mushroom Names

Most mushrooms do not have common names, only the well-known edible or poisonous species, or those with distinctive features caused our forbears to christen them. This often comes as a surprise to people used to wild flowers or birds, which nearly always have common names. Fungi, however, have always been mysterious or too poorly known to merit such names.

The scientific name of the mushroom has two Latinized words followed by the names of the authors who played a part in describing the particular species concerned. For example, if we look at *Suillus grevillei* (Klotzch) Singer we see that the name is Latinized and always in italics. The first part (*Suillus*) is the genus or group of mushrooms to which it belongs, and the second (*grevillei*) is the specific epithet, which is always in lower case letters and shows the particular species of the genus we are dealing with. Only one member of the genus *Suillus* may bear the specific name *grevillei*.

The author's names following the mushroom's name show the history of the species: who first described it and whether anyone has ever moved the mushroom to another genus (this happens to many species over the years). In this example the species was first described by Klotzch (he described it as *Boletus grevillei*), but it was later transferred to the genus *Suillus* by Rolf Singer.

Another example is *Russula lutea* (Huds. ex Fr.) S. F. Gray. Notice that the names are often abbreviated. Here Fries described the mushroom based on the concept of Hudson, but it was Gray who placed it in the genus *Russula*. Since names do change and vary from guide book to guide book I have given well-known synonyms wherever possible. Remember, the mushrooms do not change, just our ideas about them.

Pictorial Key to Major Groups

Mushrooms with gills:
Both with and without a stem –
see pages 36–134.

Boletes:
Fleshy fungi with spongy tubes on underside of cap –
see pages 16–35.

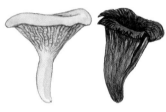

Chanterelles:
Fungi with blunt, irregular wrinkles or veins on underside of cap, or a smooth trumpet –
see pages 150–152.

Puffballs, Earthstars, and Birds Nests:
Fungi with spores inside a rounded ball, or forming tiny eggs in a 'nest'–
see pages 139–149.

Stinkhorns:
Fungi with foul-smelling spores smeared over strange fruit-bodies which hatch out of 'eggs'
– see pages 136–138.

Toothed Fungi:
Fungi with downward pointing teeth of spines, with or without a stem – see pages 160–163.

Polypores and Bracket Fungi:
Hard or tough fungi with one or more layers of downward pointing tubes – see pages 164–172.

Club and Coral Fungi:
Forming simple clubs or complex coral-like forms – see pages 153–158.
See also 183.

Jelly Fungi:
Very variable in form but all with soft, jelly-like or rubbery texture – see pages 173–175.

Morels, False Morels and Cup Fungi:
From simple cups to sponge- or brain-like structures on a hollow stem. All with spores formed in an ascus – see pages 176–185.

Development of a Mushroom with Volva and Ring

The universal veil (1) enclosing the mushroom ruptures to leave a volva (4) at the base and fragments on the cap. The partial veil (3) covers the gills (2) and then is pulled away to form a ring on the stem.

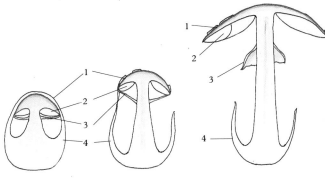

Types of Gill Attachment to the Stem and Cap Shapes

Free gills on a rounded or convex cap

Sinuate or notched gills on an umbonate cap

Adnate gills on a depressed cap

Decurrent gills on a funnel-shaped cap

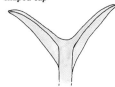

Common Mushroom Structures

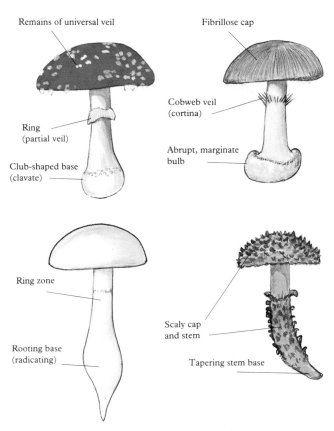

Remains of universal veil

Ring (partial veil)

Club-shaped base (clavate)

Fibrillose cap

Cobweb veil (cortina)

Abrupt, marginate bulb

Ring zone

Rooting base (radicating)

Scaly cap and stem

Tapering stem base

Spore Types

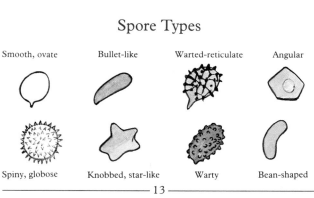

Smooth, ovate

Bullet-like

Warted-reticulate

Angular

Spiny, globose

Knobbed, star-like

Warty

Bean-shaped

Glossary

Acrid Producing burning sensation on tongue.

Adnate Refers to gills which join the stem for most of their width.

Amyloid A blue-black reaction of cells or spores to the iodine solution, Melzer's Solution.

Apiculus Point or short stalk at one end of a spore.

Basidia Club-like cells on which the spores are produced.

Caespitose Growing in clusters or clumps.

Clavate Club-shaped.

Cuticle Outer skin of the cap or stem.

Cystidium Specialized cells on any part of the mushroom, often very distinctive in shape.

Decurrent Refers to gills which run down the stem where they join.

Dextrinoid A reddish-brown reaction in cells or spores to the iodine sollution, Melzer's Solution.

Fibrillose With fine fibres.

Fibrous With coarse fibres.

Floccose Woolly or with small tufts of fibres.

Globose Roughly spherical.

Hygrophanous Changing colour as it dries out.

Hymenium A layer of cells which produce spores.

Latex The sticky fluid exuded from the tissues of some mushrooms, especially *Lactarius* and *Mycena*.

Melzer's Solution A specialized iodine solution, see Appendix for formula.

Mycelium The mass of fine threads (hyphae) which form the fungus running through the soil.

Pruinose Frosted or with a 'bloom' – like a grape.

Reticulum A fine network.

Sinuate Refers to gills which curve up or are notched where they join the stem.

Striate With fine lines.

Tomentose Finely velvety.

μm, micrometer or micron $1/1000$ of a millimeter

Umbilicate With a central depression or navel in the cap.

Umbo A central hump or nipple in the cap.

Vinaceous Wine-red, purplish-red.

Viscid Sticky, glutinous when wet.

BASIDIOMYCETES

BOLETES

GILLED MUSHROOMS

GASTEROMYCETES

CHANTERELLES

CORAL AND CLUB FUNGI

CRUST AND TOOTHED FUNGI

POLYPORES

JELLY FUNGI

Fungi with spores produced externally on cells called basidia. These basidia may be spread over gills, tubes, spines, or other external surface or contained inside the fruit-body.

BOLETES
(Boletaceae)

These fungi include some of the most sought after edible species in the world. Mushrooms such as *Boletus edulis* are eaten all over the world either fresh, dried or in mushroom soups. They share the common feature of a soft, fleshy body, the basidia produced in a vertically arranged layer of minute tubes (the pores) on the underside of the cap, and most if not all species are mycorrhizal with certain tree species. Spores vary from round to more commonly long, elliptic or bullet-like, smooth to ornamented and range in colour from yellowish to dull olive-brown, pinkish or black.

Other features to note are any colour changes, when the flesh or pores are bruised and occasionally in chemical tests, or taste and odour. In size they range from less than 2.5–60cm (1in to over 2ft) across the cap in some tropical species.

Boletus edulis
Bull. ex Fries, the Cep, or Steinpilz

Cap 8–25cm/3–10in. One of the most widely sought after edible species, it occurs over all northern temperate regions of the world. It is found in mixed conifer and deciduous forests. The young pores are white then turn yellow-olive. The flesh is unchanging when cut. Cap colour varies from yellow-brown to deep reddish-brown. The pale stem is covered in a fine network of white ridges. **Spores** 13–19 x 4–6.5µm, olive-brown.

Boletus aestivalis
Poulet ex Fries

Cap 8–20cm/3–8in. The pale, hazel-brown to ochre-brown cap is finely cracked and granular in texture. The tubes and pores are white at first then finally yellowish. The robust stem is pale brown with a white network overall while the flesh is white throughout. **Spores** 14–17 x 4.5–5.5μm. Common in some years under beech and often considered to be a better edible even than the related *B. edulis*!

Boletus aereus
Bull ex Fries

Cap 8–20cm/3–8in. The deep chocolate-brown cap, often blotched paler where leaves have covered it, and the brownish stem with darker brown reticulation are distinctive. The young pores are white then yellowish. The flesh is white. **Spores** 12–14 x 4–5μm, olive-brown. Frequent in the south of Britain and Europe, especially under oaks. A delicious edible, much sought after.

Boletus pinophilus
Pilat & Dermek

Cap 8–20cm/3–8in. A close relative of the well-known Cep bolete, *B. edulis*, this species differs in its redder, or more purplish-brown cap colours and reddish-brown stem with fine white network overall. The flesh is firm and white with a reddish flush under the cap skin. The pores are white at first then soon yellowish. **Spores** 14–17 x 4.5–5.5μm, olive-brown. Quite common under pines in Scotland and parts of England. Edible and good.

Boletus impolitus
Fries

Cap 5–15cm/2–6in. The yellowish-ochre to buffy-grey cap is finely velvety to matt. The tubes and pores are rather bright yellow but do not change colour when bruised. The stem is pale yellowish with the surface finely dotted with minute woolly flecks; the base is often slightly rooting and tapered. The flesh is firm and white or flushed slightly reddish. There is often a strong odour of iodine, especially in the flesh of the stem base. **Spores** 9–16 x 4–6µm. A rather uncommon species, usually under oaks. Edible but poor quality.

Boletus regius
Krombholtz

Cap 8–20cm/3–8in. The beautiful rose-red to dark-red cap, yellow pores which bruise blue, and pale yellow stem flushed pink are easy to recognize. The stem has a fine network overall. The flesh stains a pale blue. **Spores** 12–16.5 x 3.5–5µm. A rare but beautiful species, found in southern England, usually under oaks or beeches. Edible but often spoilt by insects.

Boletus appendiculatus
Sch.

Cap 10–20cm/4–8in. A good edible species, it has a dry, cinnamon-brown to yellow-brown cap, bright yellow pores and stem with a fine yellow network over the upper half, the base may be stained reddish brown. Both pores and flesh stain blue. **Spores** 12–15 x 3.5–5µm. Frequent under beech in central and southern England. Edible.

Boletus calopus
Fries

Cap 5–20cm/2–8in. This is one of a
number of bitter-tasting species. Its
distinguishing features are the reddish
stem with a white network, yellow-
brown to whitish-olive cap and yellow
pores which bruise blue, as does the
flesh also. **Spores** 10–14 x 4–6µm.
Under beech and conifers, quite
common throughout most of Britain.
The very bitter taste renders it
inedible.

Boletus albidus
Rocques

Cap 8–15cm/3–6in. At first dingy
white then soon dirty buff to olive-
grey, slightly velvety then smooth.
The tubes and pores are bright
lemon-yellow bruising blue. The stem
is usually fat and swollen, often
rooting, pale lemon-yellow with a very
fine network which is often reduced
or even absent. The flesh is whitish
yellow becoming very pale blue when
cut and has an extremely bitter taste.
Spores 12–16 x 4.5–6µm. Rather uncommon,
found under beech and oak in southern and
central Britain. Inedible.

Boletus erythropus
Persoon

Cap 5–15cm/2–6in. The cap
colour varies from rich
chocolate-brown to pale yellowish-
brown. The pores are usually deep,
almost blood-red but may be paler,
more orange, in colder weather. The
stem is yellowish orange overlayed
with very fine red dots (you may need
to use a hand-lens). The flesh is yellow
but instantly flushes deep blue when cut.
Spores 11–19 x 4.5–7µm. Common throughout
Britain in mixed woods. Apparently edible but
best avoided like all blue staining species.

Boletus pseudosulphureus
Kbch.

Cap 8–10cm/3–4in. The yellow cap becomes browner with age and all parts bruise deep blue when touched, the tubes and pores are yellow. The stem is yellow with slightly darker, fine dots. The flesh is bright yellow becoming blue. **Spores** 10–16 x 5–6.5µm. This very rare mushroom must be considered a prize find in Britain where it occurs under beech and oak. Inedible.

Boletus luridus
Schff. ex Fries

Cap 8–12.5cm/3–5in. The matt cap varies from pale orange-ochre to yellow-buff, becoming more olive with age. The tubes are yellow while the pore mouths are orange-red and bruise deep blue. The stem is yellowish with a prominent red network. The flesh is yellowish, often deep purple-red in the stem base, and instantly turns blue when cut. **Spores** 9–17 x 5–7µm. This species is not uncommon under beech and oak on calcareous soils, particularly in the south. Edibility doubtful, best avoided.

Boletus queletii
Schultzer

Cap 8–12.5cm/3–5in. The cap is a rich brick to orange-red while the tubes are yellow with apricot-orange pores. The pores bruise deep blue when touched. The stem is yellowish with tiny red dots and is often pointed and rooting at the base. The flesh is yellow flushed deep purple-red in the base, and turns blue when cut. **Spores** 10–17 x 5–8µm. A rather uncommon species it occurs under beech, oak and lime in southern England. Edibility uncertain, best avoided.

Boletus satanus Satan's Bolete
Lenz

Cap 8–30cm/3–12in. This often
huge, bulky bolete is easily
recognized by the swollen,
obese stem with fine red
network, and the pallid, grey-
white to olive-white cap. The
blood-red pores bruise blue
as does the pale yellow flesh.
The odour soon becomes
strong and nauseous, like
rotting garlic, and the
mushroom has caused acute
gastric upsets. **Spores**
11–15 x 4–6μm. Found under
beech and oak on calcareous
soils in central and southern
England. Poisonous.

Boletus splendidus
Martin

Cap 10–15cm/4–6in. Greyish-white to almost olive but soon
flushed with pink. The tubes are yellow with deep, blood-red
pore mouths which bruise blue. The stem is cylindrical (not
fat and swollen) and is dull yellow with a purplish-red
network especially on the upper half. The flesh is lemon-
yellow and flushes very pale blue when cut. The odour is
slightly spicy but not nauseous as in *B. satanus*. A rare species
it occurs under beech and oak in southern and central
England. Possibly poisonous.

Boletus rhodopurpureus
Smotlaka

Cap 10–20cm/4–8in. The cap is at
first grey-brown to pinkish-grey
but soon becomes flushed
purple-red overall. The tubes are
yellow with blood-red pores which bruise
blue. The stem is stout, swollen and
yellowish with a purple-red network
overall. The flesh is bright yellow and
turns bright blue when cut, there is
often a strong, fruity odour present.
Spores 14–18 x 5.5–6.5μm. A rare
species found under beech and oak on
calcareous soils in southern England.
Possibly poisonous.

Boletus subtomentosus
L. ex Fries

Cap 5–10cm/2–4in. The velvety cap is
olive-yellow to rich yellow becoming
smoother and more reddish-brown with age.
The tubes and pores are bright yellow and do
not turn blue when bruised. The stem is
narrow and tapered, dull olive-yellow to
cream-buff and often has some coarse ridges
or furrows. The flesh is white and almost
unchanging. **Spores** 11–14 x 4–6μm. A drop
of ammonia placed on the cap turns dull
brown, not blue-green. Common in deciduous woods
throughout England. Edible but poor.

Boletus lanatus
Rostkovius

Cap 2.5–10cm/1–4in. The very velvety
cap is deep yellow-buff, cinnamon,
becoming darker brown with age. The tubes
and pores are bright yellow, angular, and bruise
blue on handling. The stem is yellowish-buff
with prominent brick-coloured veins forming
a coarse network at the apex. Ammonia on
the cap turns blue-green. The flesh is white,
pale yellow in stem base and hardly blues
when cut. **Spores** 9–11.5 x 3.5–4.5μm. Common in
deciduous woods, especially birch. Edible.

Boletus chrysenteron
Fries

Cap 5–8cm/2–3in. One of the most well known species in both America and Europe, it often has a cracked cap surface with reddish flesh showing through the cracks. Cap is shades of olive-brown to reddish-brown and the pores are yellow, bruising blue. The stem is yellowish-white above shading to purplish-red below. The flesh bruises blue. **Spores** 9–13 x 3.5–5μm. Common throughout Britain in mixed woods. Edible but poor.

Boletus pruinatus
Fries

Cap 5–10cm/2–4in. The cap is a deep purple-brown, chestnut, to almost black at the centre with the margin usually red; it is covered in a hoary bloom like a grape. The tubes and pores are bright yellow, slowly blueing when bruised. The stem is smooth and bright chrome-yellow with the base flushed slightly reddish. The flesh is also bright yellow, staining slowly blue, which distinguishes it easily from the related *B. chrysenteron*. **Spores** 11.5–14.5 x 4.5–5.5μm. Quite common late in the year, under beech and oak in Southern England. Edible.

Boletus porosporus
Imler

Cap 5–8cm/2–3in. The matt, or velvety cap is a dull, olive-brown and soon becomes cracked, exposing whitish flesh beneath. The tubes and pores are olive-yellow. The stem is usually a dull greyish-olive to brown with a narrow reddish zone above and a brighter yellow above this zone. **Spores** 12–17 x 4–6.5μm and one end of each spore is distinctly flattened or truncate (this must be from a spore print, spores taken from the tubes are rarely mature and are not yet flattened). Quite common in mixed woods, especially oaks. Edible but poor.

Boletus leonis
Reid

Cap 2.5–8cm/1–3in.
The surface of the cap is
distinctly roughened to
velvety and is a rich
tawny-orange to yellow
and often slightly
cracked or scaly. The
tubes and pores are
bright yellow to olive-
yellow and do not

bruise blue. The stem is yellow-ochre
and smooth to slightly woolly. **Spores**
10–13 x 5–6µm. Rather uncommon it
occurs under oaks in southern
England. Edible.

Boletus rubellus
Krombholtz

Cap 1–5cm/½–2in. This small
bolete is a bright rose-red to scarlet
when young with a velvety surface,
becoming paler and often cracking
with age. The yellow pores and the
reddish stem both bruise blue when
handled. The base of the stem has a
distinct yellow coating. **Spores** are
brown, 7–17 x 4–7µm. Found in grass
under oaks, quite common.

Boletus rubinus
W. G. Smith

Cap 2.5–5cm/1–2in. The cap is
brownish-ochre to pale brick .
The tubes and pores are slightly
decurrent and are a deep
carmine-red. The short, tapered
stem is reddish above, yellow
below. The flesh is pale, whitish
to reddish in places. **Spores**
5.5–6.5 x 4–5µm. Uncommon,
found under oaks in grassy
areas, mainly in southern
England. Edible.

Boletus piperatus
Fries

Cap 3–8cm/1–3in. A small but attractive species with a dull, orange-brown cap and intense cinnabar-red to cinnamon-colored pores. The stem is pale brown with the base always a bright chrome-yellow, as is the flesh when the stem is split open. As the name suggests the taste is very peppery. **Spores** 9–12 x 4–5µm. Common under birch and sometimes pines throughout Britain, often in close association with *Amanita muscaria*. Inedible.

Aureoboletus cramesinus
(Secr.) Watling

Cap 2.5–8cm/1–3in. The slightly sticky cap is a dull pinkish-brown to reddish brown and often irregularly streaked. The tubes and pores are a brilliant golden-yellow and do not bruise blue. The slender, tapered stem is yellow flushed with reddish brown and often sticky. The flesh is white, marbled with pinkish brown. **Spores** 15–20 x 4–7µm. A rather uncommon species found under beech and oak on clay or calcareous soils in southern England. Edible.

Boletus parasiticus
Fries

Cap 2–8cm/1–3in. This unique species is to be found attached to specimens of the common Earthball, *Scleroderma citrinum*, whose tissues are invaded by the bolete. Often 3 or more boletes can be seen on one Earthball. The Earthball does not seem to be affected by the invading bolete and continues to produce its own spores! The pores of the bolete are often stained reddish. **Spores** 12–18 x 3.5–5µm. Locally common in damp woodlands in England, rarer in Scotland. Edible but poor.

Boletus pulverulentus
Opatowski

Cap 5–10cm/2–4in. Dull yellow-brown to reddish-brown. Pores are yellow and the yellow stem is flushed with reddish-orange below. The remarkable part of this fungus is the instantaneous change of all parts of the fruit-body to the deepest blue when bruised in any way. **Spores** 11–14 x 4.5–6μm. Uncommon, under oaks, mostly in the south and west of England. Edibility uncertain, best avoided.

Boletus badius
Fries

Cap 5–10cm/2–4in. The slightly velvety to smooth cap is a rich bay brown to yellow-brown. The yellowish pores become slowly greenish with age and bruise blue. The stem is coloured like the cap with pinkish-brown tones, often with a whitish bloom. The whitish-yellow flesh bruises pale blue. **Spores** 10–14 x 4–5μm. Very common throughout Britain in both coniferous and deciduous woodlands. Edible and good.

Boletus fragrans
Vittadini

Cap 8–12.5cm/3–5in. The cap is dark reddish-brown with matt to greasy surface while the tubes and pores are bright yellow. The stem is usually pointed and slightly rooting and is pale creamy-ochre, smooth, but darker, reddish-brown to pinkish below. The flesh is white to pale yellow, sometimes turning faintly blue when cut. The odour may be strong of fruit or spice. **Spores** 10–15 x 4.5–5.5μm. A very rare species it is found under oaks in grassy places. Edibility uncertain.

Tylopilus felleus
(Fries) Karsten

Cap 5–20cm/2–8in. Often
very large in size this appears
to be a tempting edible, and is
confused with the Cep bolete,
B. edulis, but one taste shows
the difference, it is terribly
bitter! The cap and stem are
yellowish-brown to tan while
the pores mature to deep pink.
The stem has a raised network
over the whole surface.
Spores 11–15 x 3–5μm.
Common under mixed
deciduous trees, especially
beech and oak. Inedible
because of the bitter
taste.

Uloporus lividus
(Fries) Quélet

Cap 5–10cm/2–4in.
Smooth or slightly
roughened, but sticky when
wet, the cap is pale ochre-
buff to rust-brown. The
tubes and pores are quite
deeply decurrent and deep
lemon to greenish yellow;
the pores are large and
often angular and irregular.
The stem is buff-brown
and smooth. **Spores** 4.5–6
x 3–4μm. This rare species
is only found under Alder
trees (*Alnus* spp), often in
deep grass and is easily
recognized. Edible.

Leccinum carpini
(Schulz) Pearson

Cap 5–10cm/2–4in. The genus *Leccinum* is distinguished by the woolly-squamulose stem with the squamules usually starting pale and darkening with age. The strangely wrinkled, cracked cap is a dark sepia-brown to tawny-brown becoming more cracked as it ages. The tubes and pores are white to cream bruising coral-pink then soon purplish-black. The stem is often long and slender, white with darker, sepia-brown, woolly scales. The white flesh bruises reddish-pink then soon purplish and finally black; in the stem base it is frequently bright blue. **Spores** 12–17 x 4.5–6µm. Found most commonly under hornbeam but also with hazel, usually in late summer-early autumn, widespread but especially in southern England. Edible.

Leccinum holopus
(Rostkovius) Watling

Cap 5–10cm/2–4in. This rather soft, fleshy species begins completely white with a dry to slightly sticky cap when wet. As it ages it may develop a slight greenish tinge over most of the fruit-body. The stem has small white, woolly scales which darken with age. The flesh when cut is white to olive in the stem apex, and in some forms may show a very faint flush of pink. **Spores** olive-brown, 16–19 x 4–6µm. An uncommon species, it grows in boggy areas under birch, especially in sphagnum moss and is widespread. Edible but poor.

Leccinum duriusculum
(Schulzer) Singer

Cap 7.5–12.5cm/3–5in. A large, often robust species, the cap is usually a dull grey-brown to chocolate, smooth to slightly velvety, with some wrinkles. The cuticle overhangs the edge of the cap. The pores are cream bruising brown. The robust stem is white to grey-brown with dark brown woolly scales. There is usually a flush of blue-green at the stem base. The white flesh turns grey-lilac to pinkish purple when cut. **Spores** olive-brown, 16–18 x 4–6μm. A rare species found almost exclusively under poplar trees in Britain. Edible and good.

Leccinum quercinum
(Pilát) Green & Watling

Cap 5–15cm/2–6in. The rich, fox-red to brick-red or even dark reddish-chestnut cap is minutely woolly-fibrous and may be slightly scaly at the centre, the cap cuticle overhangs the margin. The tubes and pores are cream to buff and bruise pinkish-brown. The stem is fleshy, often stout, cream with dark foxy-brown woolly squamules. The flesh is white, flushing pinkish-purple then soon purplish-grey, often greenish in stem base. **Spores** 12–15 x 3.5–5μm. Quite common in some areas, under oak, especially in central and southern England. Edible and good.

Leccinum versipelle
(Fries & Hök) Snell
Cap 8–20cm/3–8in. The bright
yellow-orange to tawny-orange cap
is minutely downy and often slightly
scaly at the centre, the cap cuticle overhangs
at the margin. The tubes and pores are cream
bruising purplish-buff. The stout stem is
white or greyish with dark, black-
brown woolly squamules, the base
of the stem is often flushed green.
The flesh is white becoming livid
purplish-pink then purplish-grey and
finally blackish, often flushed green
in the stem base. **Spores** 12.5–16 x
4–5µm. Very common and
widespread wherever birches are
found. Edible and good.

Leccinum aurantiacum
S. F. Gray
Cap 5–15cm/2–6in. The cap is a rich orange
to brick-red with the cap cuticle overhanging the
margin. The tubes and pores are pale cream
bruising pinkish-brown. The stem is cream with
woolly squamules which start white at first but
soon turn dark reddish-brown. The flesh is white
turning reddish then purplish and finally greyish-
brown. **Spores** 14–16.5 x 4–5µm. Found only under
aspens and other *Populus* species, it is widespread but
not common. Edible and good.

Leccinum crocipodium
(Letellier) Watling
Cap 5–10cm/2–4in. Starting a rich, deep yellow
the cap soon turns a dull yellowish-brown to
olive-brown. The surface is often irregular and
lumpy, becoming minutely cracked and
mosaic-like. The tubes and pores are bright
yellow bruising brownish. The stem is stout, pale
yellow with woolly squamules which darken with
age. The flesh is yellow then greyish-purple,
chestnut and finally blackish. **Spores** 12–17.5
x 4.5–6µm. Found under oak, often quite commonly
in warm, wet years in southern England. Edible.

Leccinum variicolor
Watling

Cap 5–10cm/2–4in. The often oddly blotched and mottled cap is a drab, mouse-grey, grey-brown to almost blackish-grey with whitish areas. The tubes and poores are cream bruising rose-pink. The stem is whitish with dark grey-brown to black woolly squamules on the lower half. The flesh is white turning bright rose-pink to reddish-salmon when cut (it is stronger when rubbed.) and often bright blue green in the base of the stem. **Spores** 12.5–16 x 4.5–6µm. Found under birch, this often very common species was formerly confused with *L. scabrum* which does not discolour when cut. Edible.

Leccinum scabrum
(Fries) S.F. Gray

Cap 5–15cm/2–6in. Although commonly reported, this is probably the most misidentified species of *Leccinum*. The true *L. scabrum* has a soft, dull brown to buff cap, pale whitish-buff pores that hardly stain when bruised, and stem squamules starting white then soon aging brown to black. The flesh when cut is unchanging to only very slightly pinkish-buff in the stem apex. There may be blue-green stains in the stem base. Other similar species stain a much brighter pinkish-red in the stem apex. **Spores** 15–19 x 5–7µm..Frequent under birch throughout Britain but often confused with other, reddening species. Edible.

Leccinum roseofractum
Watling

Cap 5–10cm/2–4in. The dark, blackish-brown cap may be paler, more ochre at the centre and is usually slightly greasy in texture. The tubes and pores are cream bruising pinkish-purple then brown. The stout stem is white with dark, blackish-brown scales. The flesh is white flushing pink throughout. **Spores** 15–17 x 5.5–6µm. Often quite common, it is found under birch throughout Britain. It may be confused with *L. variicolor* but this has a mottled grey-brown cap and blue-green stains in the stem. Edible.

Gyroporus castaneus
(Fries) Quelet

Cap 3–10cm/1–4in. An unusual species varying in colour from light, yellowish-brown to rich cinnamon or brick-red, sometimes with discoloured areas. The stem is coloured like the cap and is usually hollow. The pores are white to pale yellow. **Spores** 8–12.5 x 5–6µm. Rather an uncommon species it is found in warm areas under oaks, especially in southern England. Edible and very good.

Gyroporus cyanescens
(Fries) Quelet

Cap 5–12cm/2–5in. At first this species is entirely white but soon ages a dull yellowish-tan to straw colour. The surface of the cap is roughened and fibrillose as is the stem which is usually quite swollen. The most remarkable feature of this bolete however is the instant colour change to the most intense deep blue if any part is damaged. **Spores** 8–10 x 5–6µm. Uncommon, under birch and spruce, more commonly in the north of England and Scotland. Edible.

Suillus granulatus
(Fries) Kuntze

Cap 5–10cm/2–4in. This has many of
the features considered typical of the
genus *Suillus*: viscid, slimy cap when
wet, stem with small glandular dots
and a strict relationship with pines.
The cap colour varies from pale
cinnamon-brown, as shown, to quite
bright orange-brown. The pores are
pale yellow, often weeping milky
droplets when young. **Spores** 7–10 x
2.5–3.5μm. Under conifers, especially
pines, throughout Britain. Edible with
caution, it can cause diarrhoea.

Suillus luteus Slippery Jack
(Fries) S. F. Gray

Cap 5–15cm/2–6in. A common
and attractive bolete with a deep
tawny-brown to reddish-brown,
slightly streaky cap with a
glutinous, viscid coating. The stem
is noticeable for its thick,
membranous veil forming a ring at
the top, white and often flushed
with violet on the underside.
Spores 7–9 x 2.5–3μm. Found
under pines, often very common.
Edible with caution.

Suillus grevillei
(Klotzch) Singer [= *S. elegans*]

Cap 5–15cm/2–6in. An often
abundant species its cap varies
from bright yellow to deep
reddish-chestnut, while the
yellowish stem has a white,
cottony ring at the top. The pores
are yellow, bruising reddish-
brown. **Spores** 8–10 x 3–4μm.
Found only under larch, it is
widespread and often abundant.
Edible with caution, excess can
cause diarrhoea.

Suillus variegatus
(Fries) Kuntze
Cap 5–15cm/2–6in. The specific
name refers to the cap surface which
is finely hairy-velvety to squamulose at first
then becoming smooth, dry but sticky below
the tomentose layer. The colour varies from
pale yellow to orange-yellow as does the stem.
The pores are dark brownish-cinnamon
becoming dull yellow with age. The pores,
and the flesh when cut, turn blue. **Spores**
7–10 x 3–5μm. Found under conifers,
widespread and common. Edible.

Suillus aeruginascens
(Secretan) Snell
Cap 5–10cm/2–4in. The cap is a dull,
whitish-cream to dirty straw or olive with
browner areas and streaks. The tubes and
pores are large, angular, olive-buff bruising
greenish brown. The stem is straw-yellow
with a membranous ring. The flesh is cream-
straw flushed olive in the stem and staining
blue-green in the stem base. **Spores** 10–12 x
4–5μm. Found only under larches this rather
uncommon species is widespread in Britain. Edible but poor.

Suillus bovinus
(Fries) Kuntze
Cap 2.5–10cm/1–4in. The usually
sticky, smooth cap is buff to pinkish-
clay or cinnamon-brown, with the
inrolled margin distinctly paler,
almost whitish when viewed from
below. The tubes and pores are
slightly decurrent, olive-buff to ochre
and form pores within pores (referred
to as compound pores). The rather
short stem is pale ochre to brown,
minutely dotted at the apex and does
not have a ring. The flesh is whitish-
yellow staining pinkish-buff,
sometimes bluish above the tubes.
Spores 8–10 x 3–4μm. Found under
pines, often quite abundant. Edible.

Suillus cavipes

(Opat.) Smith & Thiers

Cap 5–10cm/2–4in. The cap is a dull
yellowish-red to reddish-brown, dry,
scaly-fibrous. The stem is reddish-brown
and usually hollow with a white, ring-like
zone at the apex which soon vanishes.
The pores are decurrent and dull yellow to
greenish-yellow and large, angular. **Spores**
7–10 x 3.5–4μm. Found under larch, this
is a rare species, usually in the south of
England. Edible.

Porphyrellus porphyrosporus

(Fries) Gilbert

Cap 5–10cm/2–4in. The dark olive-
brown cap is smooth to matt. The
tubes and pores are dark, grey-yellow
to grey-brown becoming slightly
purplish. The stem is smooth and
coloured like the cap. The flesh is
white turning slightly smoky-pink to
greenish below. **Spores** greyish-
purple, 10–20 x 5–10μm. An
uncommon species found under
beech and oak, sometimes conifers,
especially in the northern parts of
Britian. Edible.

Strobilomyces floccopus

(Fries) Karsten

Cap 5–10cm/2–4in. The black or greyish
cap with thick, woolly scales is quite
unmistakable. The scales usually
overhang the margin in a ragged veil.
The tubes and pores are white then grey-
olive with age and bruise red. The stem is
equally woolly with a thick ring at the top.
The flesh is white then rapidly turns first
pink then deep coral-red to orange-red
and finally reddish-brown. **Spores** are
unique in British boletes, rounded with a
coarse network, 10–12 x 8.5–11μm, and
are violet-black in deposit. A rare species
found in mixed, shady woods, widespread. Edible.

GILLED MUSHROOMS
WHITE TO PALE OCHRE SPORES

RUSSULA FAMILY (Russulaceae)

The two genera *Russula* and *Lactarius* are often placed in their own order – the Russulales – because they differ so markedly from other gilled mushrooms. They have a characteristic brittle, crumbly texture in the hand, and their spores are ornamented with a variety of warts and ridges that stain blue-black when treated with a special iodine solution (Melzer's Solution). Other important features are the taste of the cap flesh and gills, the odour, and the colour of the spore deposit. The latter is placed on a scale of pure white to ochre, designated by 8 letters (A–H) shown below and is vital to accurate identification, especially in *Russula*.

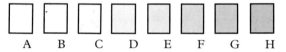

| A | B | C | D | E | F | G | H |

Lactarius differs from *Russula* in its usually duller colours and in its flesh which oozes a sticky latex when cut. This latex may be clear, white, or coloured. The spore colour is less variable in *Lactarius*. Both groups are very large and can be very difficult to distinguish. Only the most striking and easily identified species are shown here.

Russula nigricans
Fries

Cap 5–15cm/2–6in. The cap starts white and rounded but soons ages blackish-brown and becomes funnel-shaped. The cap cuticle does not peel off easily. The gills are very thick and very widely spaced with a number of shorter, intermediate gills present. The stem is short and firm, coloured like the cap. The flesh is white then soon red and finally black after cutting, and has a mild flavour. **Spores** white (A), 6–8 x 6–7μm, with a fine network. Common in mixed woods throughout Britain. Edibility doubtful.

Russula albonigra
Fries

Cap 5–10cm/2–4in. This all-white
species soon turns black where
bruised or with age. It does not pass
through a reddish stage as in the
previous species. The gills are rather
crowded, cream-buff. The taste is
mild with a menthol-like aftertaste. It
is quite frequent under beech. **Spores**
white (A–B), 7–10 x 6–8μm, with low
warts and a partial network. Inedible.

Russula delica
Fries

Cap 5–15cm/2–6in. The all white
cap is soon funnel-shaped, often
pitted or wrinkled on the surface.
The crowded gills are white or often
flushed faintly blue-green. The short,
firm stem is white as is the flesh which
does not change colour when cut. The
odour is fruity then fishy and the taste is
mild to peppery. **Spores** 8–11.5 x 6.5–
8.5μm, with short warts and a partial
network. Common throughout Britain under deciduous trees.
Edible but poor. The variety with the blue-green tint is often
referred to as a separate species *R. chloroides.*

Russula cyanoxantha
(Sch.) Fries

Cap 5–10cm/2–4in. The cap colour
varies from pure lavender to pure
green or mixtures of both, the
surface is often finely veined or
radially streaked and the cuticle
peels about half way to the centre.
The gills are quite crowded and usually
unforked although this varies; when
rubbed the gills are soft, flexible and
greasy. The often tall stem is firm and
white. The taste is mild to slightly
sharp. **Spores** white (A) 6.5–10 x
5.5–6.5μm, with small isolated warts. A very common species
throughout Britain in mixed woods. Edible.

Russula ionochlora
Romagnesi

Cap 5–10cm/2–4in. The rather matt, dry cap is usually a mixture of lavender, grey and green or brown and any of these colours may dominate although lavender is the most common. The gills are pale cream and are brittle to the touch (compare with *R. cyanoxantha*). The stem is white and the taste is mild. **Spores** pale cream (C), 6.5–7.5 x 5–6µm, with small, isolated warts. Often very common under oaks and beech in parts of southern England. Edible.

Russula parazurea
Schaeffer

Cap 5–10cm/2–4in. The matt, grey-blue to grey-brown or lilac cap is usually strongly pruinose with a thick bloom like a grape especially at the margin. The gills are pale cream and brittle. The stem is white or sometimes flushed violet. The flesh is mild. **Spores** pale cream (B–C), 6–8.5 x 5–6.5µm, with low warts and a fine network. Frequent under oaks and limes in England. Edible.

Russula virescens
(Sch.) Fr.

Cap 5–10cm/2–4in. This attractive species varies from pale blue-green to pale yellow-green or ochre-brown but the cap surface is always cracked into small woolly patches. The gills are white, rather widely spaced. The stem is firm and white and the flesh is mild to taste. **Spores** white (A), 6–9 x 5.5–7µm, with isolated warts or sometimes a few connecting lines. Quite a common species under beech and oak, widespread. Edible and good.

Russula aeruginea
Lindblad

Cap 5–10cm/2–4in. Varying from pure green to yellowish or greyish-green, often with slightly rusty spotting. The taste is mild. **Spores** deep cream (D–E), 6–10 x 5–7μm with low warts and few connecting lines. A common species in mixed woods throughout Britain. Edible.

Russula emetica
(Sch.) Persoon

Cap 5–8cm/2–3in. The bright scarlet, cherry-red cap is smooth and sticky when wet, the colour may fade slightly when old. The gills and stem are pure white, soft and fragile. The taste of the flesh is extremely hot and the odour is faint of coconut. **Spores** 7.5–12.5 x 6–9.5μm, with tall warts and a well-developed network. Found in wet, boggy woods, frequent in sphagnum moss under pines. Inedible.

Russula mairei
Singer

Cap 5–10cm/2–4in. The bright red, rather matt cap may wash out after rain to a paler cream-pink, the cuticle peels very easily and almost entirely off the cap. The gills are white with a very faint blue-green hue when young, more cream with age. The stem is white and the flesh is very acrid to taste. There is often an odour rather like honey in older specimens. **Spores** white (A), 7–8 x 6–6.5μm, with medium warts in a well-developed network. Very common under beech trees everywhere. Inedible.

Russula betularum
Hora

Cap 2.5–5cm/1–2in. This
pretty little species has a
delicate pale rose to almost
whitish cap, snow-white
gills and stem but a fiercely
hot taste. Inedible. It is not
uncommon in damp
woodlands especially under
birch. **Spores** white (A),
8–10.5 x 6.5–8μm with
warts connected by a
partial network.

Russula fragilis
Fries

Cap 1.5–5cm/1/2–2in. The
cap is variable in colour
usually in mixtures of purple
and green, and very fragile.
The gills are white and
usually minutely serrated.
The stem and flesh are
white and the taste is very
acrid, burning. **Spores** white
(A–B), 6–9 x 5–8μm, with a
well-developed network.
Common throughout Britain
in mixed woods. Inedible.

Russula lutea
(Hudson) Fries

Cap 5–8cm/2–3in. The
delicately coloured rosy-
peach cap contrasts with
the deep ochre gills and
white stem. The cap cuticle
peels completely and the
taste is mild. **Spores** deep
ochre (G–H), 7–9 x
6–8μm, with rather tall,
isolated warts. A rather
common species under
beech and oak. Edible.

Russula puellaris
Fries

Cap 2.5–5cm/1–2in. A small, fragile species, the dull reddish-brown to purplish-brown cap discolours dull yellow as do the gills and stem all over in about 2 hours. The taste is mild and the cap peels about two-thirds to the centre. **Spores** pale ochre (D–E),6.5–9 x 5.5–7μm, with rather tall, isolated warts. Quite common in damp woods throughout Britain. Edible.

Russula vinosa
Lindblad = *R. obscura*

Cap 5–12.5cm/2–5in. The cap is smooth, deep purple to wine-red, often depressed in the centre. The gills are broad, pale ochre and discolouring grey-black with age. The tall stem is white but also discolours grey-black as it ages. The taste is mild. **Spores** medium ochre (D–E), 8–11 x 8–9μm, with large, isolated warts. A rather rare species, found in wet conifer woods, especially in Scotland.

Russula claroflava
Grove

Cap 5–10cm/2–4in. The vivid yellow cap, cream gills and slowly blackening flesh characterize this species. The taste is mild and it is frequent in wet, swampy areas under birch. **Spores** pale ochre (E–F), 7.5–10 x 6–7.5μm, with more or less isolated warts. A good edible species.

Russula xerampelina

(Sch.) Fries = *R. erythropoda*
Cap 5–10cm/2–4in. The cap
is a deep purple to wine-red,
sometimes almost black or
olive-black at the centre, the
surface is dry and slightly
wrinkled concentrically. The
gills are pale ochre while the
stem is a beautiful purplish-
red or pink but stains yellow-
brown on handling. The flesh
stains brown when cut and as
it ages smells strongly of old
fish or crab. The taste is mild.
Spores pale ochre (E–F),
7.5–10 x 7–8.5μm, with tall,
isolated spines. The flesh with
$FeSO_4$ gives a deep green
reaction. Rather uncommon,
found in pine woods
throughout Britain. Edible.
There are several related
species, differing in colour
(usually their stems are white)
but all have the fishy odour
and all give a deep green
reaction to $FeSO_4$.

Russula olivacea

(Schaeffer) Fries
Cap 10–20cm/4–8in. This
magnificent species
usually has a purplish cap,
despite the specific name,
but will occasionally be
flushed olive. The gills are
deep ochre and the white
stem is flushed pink,
especially at the apex. The
taste is mild. A frequent
species in some areas
under beech. **Spores** deep
ochre (G–H), 8–11 x
7–9μm, with very tall,
isolated warts.

Russula atropurpurea
Krombholtz

Cap 5–10cm/2–4in. The rich purple-red cap may be almost black at the centre and is sometimes irregularly blotched with paler areas. The cream gills are often spotted rust-red. The stem is white, sometimes flushed grey with age. The taste varies from mild to quite peppery. **Spores** white (A), 7–9.5 x 6–8μm with a fine partial network. Very common in mixed woods, especially under oaks, throughout Britain. Edible.

Russula caerulea
Fries

Cap 5–8cm/2–3in. One of the very few species with a distinctly umbonate cap, it is a deep purple-violet. The gills are broad, ochre. The stem is white, club-shaped and the flesh is white and mild to slightly bitter to taste. **Spores** deep ochre (F–G), 8–10 x 7–9μm, with tall warts forming short ridges. Quite common under pines throughout Britain. Edible.

Russula vesca
Fries

Cap 5–10cm/2–4in. The cap is usually a mixture of pinkish-browns to purplish-brown and the cap cuticle peels about half way to the cap centre. The cuticle at the cap margin often retracts to reveal a narrow strip of flesh at the edge.. The gills are cream, rather crowded and often forking near the stem. The stem is white and the flesh is mild to taste. When a drop of $FeSO_4$ is placed on the stem a deep salmon-brown reaction occurs. **Spores** white (A–B), 6–8 x 5–6μm, with small warts, mostly isolated. Common in deciduous woods throughout Britain. Edible.

Russula amoena
Quélet

Cap 5–8cm/2–3in. The cap is a beautiful reddish-purple to violet and has a velvety-matt texture. The gills are dark cream, often with the outer edges reddish. The stem is white flushed with pinkish-purple and the flesh is mild to taste. The odour of the flesh is like that of Jerusalem artichokes. **Spores** cream (B–C), 6–8 x 5.5–7µm, with short warts and a partial network. Rather uncommon under mixed woods in England. Edible.

Russula violeipes
Quélet

Cap 5–8cm/2–3in. The colour varies from pale lemon-yellow (the commonest hue) to greenish-yellow or even flushed purple, the cuticle is dry and matt. The gills are pale cream. The stem is usually flushed with violet-red or purple tones and develops an odour rather like crab. **Spores** cream (B–C), 6.5–9 x 6.5–8µm, with low warts in a partial network. Quite frequent under deciduous trees in Southern England. Edible.

Russula laurocerasi
Melzer

Cap 5–10cm/2–4in. The often very sticky, viscid cap is a pale honey-brown with a coarsely ridged and pimpled margin. The gills are cream, speckled reddish-brown. The often tall stem is white to pale buff and the flesh varies from mild to quite hot and unpleasant to taste. The odour is remarkable however, strong and fragrant of bitter almonds (marzipan) with only occasionally a sour undertone. **Spores** cream (B–C), 7–8.5 x 6.5–7µm, with remarkably tall (2µm) ridges and flanges sticking out like wings. Not uncommon under oaks, widespread. Not edible.

Russula foetens
Fries

Cap 8–15cm/3–6in. The often large cap is a dull tawny-brown to honey-brown, often very sticky-viscid and with coarse ridges on the margin. The gills are cream, spotted with rust-red and crowded. The stout stem is cream stained brown at the base, rather rigid and brittle and often hollowed out inside. The flesh is mild to unpleasant in the stem but the gills are very acrid, while the odour is sour like rancid cheese or very oily. **Spores** cream (B–D), 8–10 x 7–9μm, with tall, isolated warts. Quite common under mixed trees throughout Britain, often only just pushing through the soil. Not edible.

Russula amoenolens
Romagnesi

Cap 5–10cm/2–4in. The rather dull, sepia-brown cap with strong marginal grooves, the fetid, cheesy odour combined with an unpleasant, hot taste are all distinctive. It grows commonly under deciduous trees. **Spores** cream (B–D), 6–9 x 5–7μm, with mostly isolated warts. Inedible.

Russula farinipes
Romell

Cap 5–8cm/2–3in. The rather bright yellow-ochre cap is matt with a rather elastic cuticle and is finely furrowed and ridged at the margin. The gills are cream, rather distant, very flexible and elastic to touch. The stem is white to pale ochre with the upper portion very mealy-granular. The flesh is white with a rather hot taste and the odour is a little like fruit. **Spores** pure white (A), 6–8 x 5–7μm, with isolated warts. Under deciduous trees on clay soils, very uncommon. Edibility doubtful.

Russula fellea
Fries

Cap 5–10cm/2–4in. The cap, gills and stem are all honey-yellow to deep ochre. The mushroom, particularly the gills, has a distinct odour of pelargonium (household geraniums). The taste is very acrid. **Spores** cream (B–C), 7.5–9 x 5–6μm, with tall warts in a partial reticulum. Very common under beech, widespread. Inedible.

Lactarius deliciosus
(Fr.) S.F. Gray

Cap 5–10cm/2–4in. The orange, zonate cap stains green with age as do the gills and stem. The stem is usually pitted with small spots. When cut the flesh oozes bright orange latex. The taste is mild. Often abundant under pines it is an edible species. **Spores** cream 7–10 x 6–7μm, with a fine network.

Lactarius salmonicolor
Heim & Lecl.

Cap 5–10cm/2–4in. The entire mushroom cap, gills and stem is a bright orange. The cap is concentrically zoned and the stem may be slightly pitted with darker spots. The latex is also orange, becoming darker, reddish after a few minutes and has a soapy, bitter taste. **Spores** cream, 9–12 x 6.6–7.5μm, with a partial network. A rather rare but very beautiful species found under fir (Abies). Edible. Distinguished from the better known *L. deliciosus* by the complete lack of green stains and the tree association.

Lactarius turpis
(Weinm.) Fries

Cap 8–15cm/3–5in. This rather ugly mushroom has a deep olive-brown to blackish-green cap which is rough or matt and often concentrically zoned. The crowded gills are greenish-white and when cut ooze a white milk which stains olive. The short stem is coloured like the cap and is often pitted. **Spores** cream, 6–8 x 5.5–6.5μm, banded with ridges. Very common under birch, sometimes pine, throughout Britain. Inedible as the taste is both bitter and extremely unpleasant. Ammonia placed on the cap produces a deep violet reaction.

Lactarius vietus
(Fr.) Fries

Cap 2.5–8cm/1–3in. The cap, gills and stem are a pale lilac-grey to brownish-grey or brownish-violet. The latex is white soon staining dark grey. The taste is mild to slightly acrid. **Spores** cream, 7–8 x 5.5–6.5μm, with low warts in a partial network. Common under birch especially in Sphagnum, throughout Britain. Inedible.

Lactarius torminosus.
(Fr.) S.F. Gray

Cap 5–10cm/2–4in. The pale pink cap is zonate with a very hairy margin. The latex is white, unchanging and painfully acrid. This species is usually regarded as poisonous. Frequent under birch throughout Britain. **Spores** cream, 7.5–10 x 6–7.5μm with a network. The similar *L. pubescens* is paler, less zonate and less shaggy, and is found in similar areas.

Lactarius helvus
(Fr.) Fries

Cap 5–15cm/2–6in. The cap
is dry, slightly roughened and
pale cinnamon-brown. The
latex is watery, mild. The
odour is distinctive, variously
described as chicory,
burnt–sugar or curry. A
common species in mixed
woods throughout Britain.
Edibility is doubtful and best
avoided. **Spores** buff, 6–9 x
5.5–7.5μm, with mostly
isolated warts.

Lactarius rufus
(Fr.)Fr.

Cap 5–10cm/2–4in. The dry,
bay-red to brick-red cap usually
has a central knob. The latex is
white, abundant, and slowly very
acrid. The brownish stem has a
white base. A common species in
boggy areas under pine and birch
across the country it is usually
regarded as inedible. **Spores**
cream, 7.5–10.5 x 5–7.5μm with
a network.

Lactarius volemus
Fries

Cap 5–10cm/2–4in. The orange-
brown cap is dry and almost suede-
like to touch with a characteristically
wrinkled, puckered surface. The
rather crowded gills are creamy-
yellowish and when cut bleed
copious amounts of white, sticky
latex. This rapidly discolours brown.
The taste is mild while the odour is of
fish. **Spores** 7.5–10μm, almost globose,
with a fine network. A rather rare
species in Britain where it is found
under oaks. Edible and good.

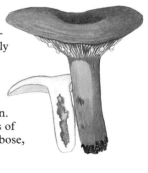

Lactarius pyrogalus
(Pers.) Fries

Cap 5–10cm/2–4in. The smooth, dry
cap is a dull ochre-brown to olive-
ochre, or sometimes greyish-ochre,
often with concentric zones. The gills
are a deep orange-ochre, widely
spaced. The stem is coloured like the
cap. The latex is white and burningly
acrid to taste (the latin specific name
translates as 'fire-milk'). **Spores**
6.5–7.5 x 5–6µm, with ridges and
bands. Common under hornbeam
and hazel throughout Britain. Inedible.

Lactarius lignyotus
Fries

Cap 2.5–10cm/1–4in. The
blackish-brown cap is velvety and
slightly wrinkled contrasting with
the almost white gills. The latex is
white, abundant, and soon
discolours the damaged tissues
bright rose. The taste is slightly
peppery. **Spores** bright ochre,
9–10.5 x 9–10µm, with a network.
Found under conifers, especially
spruce. Edible.

Lactarius camphoratus
(Fr.) Fr.

Cap 2.5–5cm/1–2in. The deep
reddish-brown to liver–coloured
cap has a small knob at the
centre while the gills and stem
are cinnamon to ochre-brown.
The latex is watery white,
scanty and mild. As it dries the
mushroom gives off a powerful
scent reminiscent of curry
powder, burnt sugar or chicory.
It is very common in
throughout Britain. **Spores**
yellowish, 7–8.5 x 6–7.5µm,
with a few connecting lines.

Lactarius vellereus
(Fr.) Fries

Cap 8–20cm/3–8in. This often
huge species has a white, funnel-
shaped cap when mature, while in
the young caps the margin is
tightly rolled in. The cap
surface is dry and velvety and
often cracks in hot, dry weather.
The thick gills are quite widely
spaced, and coloured pale
yellowish-cream. The short, hard
stem is white and velvety. The
latex is white, often copius and
very acrid to taste and very sticky
to touch. **Spores** buff, 9–12 x
7.5–10µm, with very fine ridges.
Very common under deciduous
trees throughout Britain. Inedible.

Lactarius piperatus
(Fr.) S. F. Gray

Cap 5–15cm/2–6in. The white
cap when young is rounded
then soon funnel-shaped, dry
and slightly velvety. The gills
are extremely crowded, so
much so that you can hardly see
between them, and are coloured
white to cream. The latex is
often copious, white and very
acrid to taste. In some forms it
dries olive-green. The stem
length is about the same as the
cap diameter and is coloured
like the cap. **Spores** white, 6–8
x 5–5.5µm, with a faint
network. This inedible species is
frequent under deciduous trees
throughout Britain, especially
on calcareous soils.

Lactarius tabidus
Fries.

Cap 2.5–8cm/1–3in. The pinkish-ochre to brick or cinnamon cap has a sharp umbo at the centre with tiny wrinkles around it. dries it becomes much paler, more yellow-buff. The gills are pale buff and rather crowded. The slender, smooth stem is pinkish-cinnamon like the cap. The latex is white, but slowly stains yellow when dabbed on a white handkerchief. The taste of the latex is mild and rather scanty. **Spores** cream, 7–9 x 6–7.5μm, with tall ridges forming a partial network. This is a very common species throughout Britain in deciduous woods. Edible but of poor quality.

Lactarius chrysorheus
Fries

Cap 2–8cm/1–3in. The pale, yellowish-cream to slightly pinkish cap is often marked with concentric zones or watery spots. The surface may be rather sticky when wet. The gills are moderately spaced, pale cream to buff. The latex is quite abundant, white but turns rapidly bright yellow on the gills. The flesh also turns yellow when cut. The taste of the latex is mild. **Spores** pale yellow, 6–8 x 5.5–6.5μm, with isolated warts and a slight network. This is a common species throughout Britain found under oaks. Edible.

Lactarius glyciosmus
(Fr.) Fries
Cap 3–8cm/1–3in. The entire
mushroom is pale grey with a hint of
pink or lavender but the most
distinctive character is the pleasant
odour of dried coconut. The latex is
white, unchanging and tastes very
slightly acrid. **Spores** cream, 6–8 x
5–6μm, with bands and ridges
forming a broken network. A
common species, it is edible but of
poor quality.

Lactarius blennius
(Fr.) Fries
Cap 5–10cm/2–4in. The often
very glutinous, viscid cap is a
deep olive-brown to grey-
green, usually with darker
spots. The gills are crowded,
white to spotted olive-grey.
The stem is coloured like the
cap, also viscid. The latex is
white, acrid, staining olive-
grey. **Spores** 7.5–8 x 5–6μm,
with a partial network.
Common under beech
throughout Britain. Inedible.

Lactaris quietus
(Fr.) Fries
Cap 5–10cm/2–4in. The dull,
reddish-brown cap is usually zoned
with darker bands and the gills and
stem are a similar colour. The latex is
white to slightly yellowish, mild then
soon slightly acrid. The odour is
distinctive but hard to describe, most
people say it is oily-sweet. **Spores**
cream, 7–9 x 5.5–7μm, with mostly
isolated warts and ridges. A common
species found only under oak,
throughout Britain. It is edible but
poor.

HYGROPHORUS FAMILY

(Hygrophoraceae)

All these mushrooms share the common character of thick, waxy gills, and smooth white spores formed on very long basidia. *Hygrophorus* species are thicker, fleshier and usually duller coloured than the fragile, often brilliantly coloured *Hygrocybe*.

Hygrocybe pratensis
Pers.

Cap 5–10cm/2–4in. The cap soon becomes flattened, often with a central boss and the gills run down the stem. The whole fungus is pale orange-buff. **Spores** white, 6–7 x 4–5μm. A common species it occurs in grassy areas in woods and fields. Edible and quite good.

Hygrocybe virginea
(Wulf. ex Fr.) Orton & Watling (= *H. niveus*)

Cap 2.5–5cm/1–2in. The entire mushroom is pure white, smooth to slightly greasy, hygrophanous, with a translucent, striate cap margin when wet. The gills are distant and run down the stem (decurrent). **Spores** white, 7–10 x 5–6μm. A common species in fields and pastures throughout Britain. Edible.

Hygrocybe conica
(Scop.) Kumm.

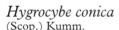

Cap 2.5–5cm/1–2in. The often acutely conical cap is bright red while the stem is yellowish. The gills are pallid to yellow. All parts of the fungus stain black on handling and after a few hours the entire mushroom will be blackened. **Spores** white, 8–12 x 5–7μm. A common species in woodlands and grassy clearings throughout Britain.

Hygrocybe punicea
(Fr.) Kummer
Cap 5–10cm/2–4in. This large, beautiful species is a bright scarlet to blood-red with yellowish-orange gills. The base of the stem and the flesh inside are white.
Spores white, 8–11 x 5–6μm. Widespread in fields and grassy clearings in woods. Edible.

Hygrocybe miniata
(Fr.) Kummer
Cap 2.5–5cm/1–2in. The brilliant orange-red to scarlet cap is minutely scurfy-scaly at the centre. The gills are orange-red and broadly attached to slightly decurrent. **Spores** white, 7.5–11 x 5–6μm. A frequent species in fields and pastures, especially in northern Britain. Edible.

Hygrocybe psittacina
(Schff. ex Fr.) Wuensche
Cap 1–2.5cm/½–1in. This remarkable species is rightly called the Parrot because its cap varies from intense green to blue-green, yellow to orange and can change within 2 hours. The surface is always very glutinous on both cap and stem. The gills are greenish to yellow and adnate. **Spores** white, 8–10 x 4–5μm. A common species in grassy clearings and woodlands throughout Britain. Edible but worthless.

Hygrocybe laeta
(Pers. ex Fr.) Karsten
Cap 1–3cm/½–1in. The
slimy-glutinous cap varies
from orange-brown to pinkish-
brown while the stem often
has a greenish or greyish apex
and is also slimy. The gills are
decurrent and greyish–pink.
Spores white, 6–8 x 4–5μm.
A widely distributed species in
woods and clearings
throughout Britain.
Edible but poor.

Hygrocybe unguinosa
(Fr.) Karsten
Cap 2.5–5cm/1–2in.
Unmistakable with its
entirely grey, very viscid-
glutinous cap and stem. The
gills are thick, deeply adnate
and also grey. **Spores** white,
6–8 x 4–6μm. Frequent in
fields, lawns and pastures
throughout Britain. Edible
but worthless.

Hygrocybe ovina
(Bull. ex Fr.) Kuehn.
Cap 2.5–8cm/1–3in. This
sombre, dark grey-black to
brownish-black mushroom
bruises reddish wherever it
is handled or broken and
often has an odour of
ammonia or bleach.
Spores white, 7–12 x
4.5–6μm. An uncommon
species it is difficult to spot
against the dark leaf-litter
in which it likes to grow, in
deciduous woods mainly in
northern Britain
Edibility doubtful.

Hygrophorus chrysodon

(Batsch) Fr.

Cap 3–8cm/1–3in. This all-white fungus is distinctive for its stem apex and cap margin are speckled with bright golden-yellow flecks. The white gills run down the stem. The smell is faint, said to resemble Jerusalem artichokes. **Spores** white, 8–10 x 4–5µm. An uncommon species, found under beech and oak. Edible but poor.

Hygrophorus penarius

Fries

Cap 10–15cm/4–6in. One of the largest of the white *Hygrophorus* species, the thick, creamy gills are widely spaced and run down the stem a little. The cap is dry to slightly downy to touch. **Spores** white, 6–8 x 4–5µm. Found under beech trees. A rather rare species, mostly in southern England. Edible.

Hygrophorus russula

(Sch.) Quel.

Cap 8–15cm/3–6in. The specific name refers to its resemblance to mushrooms of the genus *Russula*, and it does indeed look like one with its reddish cap and short, squat stature. The cap is usually speckled with darker, reddish-brown to wine-red spots on a paler background. The rather crowded gills are pinkish-cream, also often spotted wine-red. **Spores** white, 6–8 x 4–6µm. A very distinctive but rare species in Britain, found under beech on calcareous soils.

Hygrophorus hypothejus
(Fr.)Fries

Cap 5–10cm/2–4in. This very
glutinous, viscid species has a
yellowish-brown to olive-brown cap,
yellowish gills and stem with a slight
annular zone at the stem apex.
Spores white, 8–9 x 4–5μm. A
common species under conifers
particularly at the end of the year,
often after the first frosts, widely
distributed throughout Britain. Edible
but poor quality.

PLEUROTUS FAMILY

(Pleurotaceae)

These genera and species all grow on wood and usually
have the stem reduced or completely absent. The spores
vary from white to pale lavender to faintly pinkish.

Pleurotus ostreatus   Oyster Caps
(Jacq.) Kumm.

Cap 5–15cm/2–6in. The caps
are semicircular and broadly
attached at the rear to the wood
from which they grow. They
frequently grow in large
numbers, forming overlapping
clusters. The colour varies from
deep bluish-black to pale grey-
brown. The gills are cream,
narrow and very crowded.
Spores pale lilac, 7–11 x
3–4μm. On dead or dying trees,
usually late in the year, often
even in frosty weather.
Common and a good edible
species. The equally common
P. pulmonarius occurs much
earlier in the season and
is much paler, often almost white.

Pleurotus cornucopiae

Paul ex Fr. (= *P. sapidus*)
Cap 5–15cm/2–6in. This
species is unusual in having an
almost central stem and
forming a more funnel-shaped
structure. The overall colour is
pale buff to cream. The gills are
deeply decurrent and tend to
form a mesh at the base.
Spores lilac, 8–11 x 3.5–5µm.
Frequent on dead deciduous
trees and is widely distributed.
Edible and good.

Phyllotopsis nidulans

(Pers. ex Fr.) Singer
Cap 2.5–8cm/1–3in. The
bright yellow-orange cap is
semicircular to kidney-shaped
and hairy. The gills are
crowded and a deeper shade
of yellow-orange. The odour is
strong and unpleasant, almost
fetid. **Spores** pale yellowish
pink, 4–5 x 2–3µm. On fallen
conifers, often in winter or
spinrg, uncommon. Not
edible.

Panus tigrinus

(Bull. ex Fr.) Singer
Cap 2.5–8cm/1–3in. The
domed cap is almost
white with small blackish
scales, while the gills are
cream, minutely serrated
and run down the stem.
Spores white, 7–8 x
3–3.5µm. An uncommon
species on fallen timber of
poplar and willow, in
southern England. Edible
but tough.

Panellus serotinus
(Hoffm.) Kuhn.
Cap 5–10cm/2–4in. The cap
is broadly attached at the rear
or sometimes has a very short
stem, the colour varies from
deep olive-green to brownish-
yellow, often with a violet
flush and the surface is
slightly velvety. The gills are
pale yellow-orange and
crowded. **Spores** yellowish,
4–6 x 1–2µm. Quite common
on fallen timber, especially
late in the year, throughout
Britain. Edible.

Panellus stypticus
(Bull.) Karsten
Cap 1–2.5cm/½–1in. The
shell-shaped cap has a
small but distinct stem at
one edge. The surface of
the cap is dull ochre-buff
and minutely velvety. The
gills are crowded, pale buff
and often forked. **Spores**
white, 3–5 x 1.5–3µm. A
common species on fallen
deciduous wood,
especially late in the
season. Inedible.

TRICHOLOMA FAMILY

(Tricholomataceae)

An enormous family of mushrooms varying greatly in appearance and very difficult to characterize. They all share white to pale pinkish spores. Some (*Clitocybe*) have decurrent gills, others (*Tricholoma*) have sinuate gills. They range in size from tiny to huge, and some genera have veils.

Omphalina pyxidata
(Bull.) Quel.
Cap 0.6–2.5cm/¼–1in. The funnel-shaped cap is rusty-brown with a fluted margin. The decurrent gills are slightly paler as is the rather short stem. **Spores** white, 7–10 x 4.5–6µm. A common species in grassy woodlands, lawns everywhere. Edibility unknown.

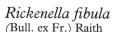

Rickenella fibula
(Bull. ex Fr.) Raith
Cap 0.6–1cm/¼–½in, depressed at centre and fluted on margin, bright orange-brown to reddish-brown. Gills are widely spaced and decurrent. The stem is slender and rather tall in proportion to the cap. **Spores** white, 4–5 x 2–2.5µm. A common species in mossy areas in woods and fields, widely distributed.

Cantharellula umbonata
(Gmel. ex Fr.) Singer
Cap 2.5–5cm/1–2in. Soon depressed at the centre with a central umbo the cap is grey to blackish-grey and often has a violet flush. The gills are paler, cream, rather thick and strongly forking. The stem is tall and colored like the cap. The flesh when damaged turns reddish-brown. **Spores** white, 7–11 x 3–4µm. A rather rare species, found in mossy areas in mixed woods. Edibility doubtful, best avoided.

Xeromphalina campanella
(Batsch. ex Fr.) R. Maire
Cap 0.6–1cm/¼–½in. This
species grows in enormous
numbers forming beautiful
drifts over decaying conifer
stumps and logs. The caps are
bright golden orange while the
decurrent gills are yellow.
Spores white, 5–8 x 3–4µm.
On dead coniferous wood, rare,
mostly in Scotland.

Armillaria mellea Honey Fungus
(Vahl.) Kumm.
Cap 5–10cm/2–4in. The
honey-yellow to greenish cap
is almost smooth or may
have very fine, brownish
scales at the centre. The gills
are adnate to slightly
decurrent, creamy-white to
pinkish. The tough, fibrous
stem is swollen, pale honey-
yellow to greenish-buff and
often with a yellow, woolly
coating at the tapered base.
There is a well-developed
woolly ring above, white to
yellow. **Spores** white, 7–10
x 5–6µm, smooth. Often in
huge clumps on dead or
dying trees, if the bark is
peeled back then long, very
tough, black 'bootlaces' may
be found. These
rhizomorphs are the Honey
Mushroom's way of
spreading long distances to
infect other trees, it is a
deadly parasite of forest
trees. Edible and good when
cooked well.

Armillaria bulbosa Honey Fungus
(Barla) Romagn.

Cap 5–10cm/2–4in. The pale
reddish-brown to pinkish-brown caps
have darker brown scales at the centre
and remains of white veil at the
margin. The gills are cream to pale
pinkish-buff. The swollen, club-
shaped stems are pale pinkish-buff
with a white, cobwebby veil above
and often a yellowish coating at the
base. **Spores** white, 7.5–8.5 x
4.5–5µm, smooth. Often in large
numbers but may be scattered over a
large area, usually on the ground on buried wood or
roots of dying trees. Edible when well cooked.

Armillaria ostoyae
Honey Fungus
Romagn.

Cap 5–10cm/2–4in. The caps are
pale to dark reddish-brown with dark,
blackish-brown scales. The gills are
cream to pale pinkish-buff. The stems
are tapered, pale brown with a white,
woolly ring that has dark brown scales
on the edge and undersurface. **Spores**
white, 8–10 x 5–6µm, smooth.
Common, in large clumps on dead
or dying timber, widely distributed.
Edible when well cooked.

Armillaria tabescens
(Scop. ex Fr.) Emel.

Cap 5–8cm/2–3in. The ochre-
brown to honey-yellow caps are
smooth with tiny brown scales at the
centre. The gills are adnate to slightly
decurrent and pale pinkish-buff. The
slender stem is completely free of any
ring or veil. **Spores** white, 8–10 x
5–7µm. Often growing in enormous
clusters at the base of oak and other
deciduous trees, in summer and early
autumn. Edible.

Hygrophoropsis aurantiaca False Chanterelle
Maire (Wulf. ex Fr.)

Cap 2.5–8cm/1–3in. This species is
white-spored but has a similarly soft,
inrolled, cap margin like *Paxillus*. Colour
ranges from pale yellow to orange while
the decurrent, forked and rather blunt
gills are bright orange. **Spores** 5–8 x
3–4.5μm. It grows under pine and birch
in damp locations throughout Britain.
It is sometimes picked in mistake for
the edible Chanterelle. It is not
poisonous, just of poor
quality.

Clitocybe clavipes
(Pers. ex Fr.) Kummer

Cap 5–8cm/2–3in. The dull grey-
brown cap contrasts with the pale
yellowish-cream, deeply decurrent
gills. The greyish stem is swollen,
club-shaped at the base. This
mushroom often has a sweet, fragrant
odour. **Spores** white, 6–8.5 x 3–5μm.
A common species in mixed woods.
Edibility suspect, reported to cause
poisoning when consumed with alcohol.

Clitocybe geotropa
(Bull. ex Fr.) Quel.

Cap 5–20cm/2–8in. This
often very large species soon
becomes funnel-shaped with
a central umbo. The colour
ranges from dull ivory-white
to ochre and the cap surface
is finely roughened. The
decurrent gills are pale
cream and crowded. **Spores**
white, 6–7 x 5–6μm.
Frequent under deciduous
trees on calcareous soils.
Edible but best avoided in
case of confusion with other
less edible species.

Clitocybe gibba

(Pers. ex Fr.) Kummer =
C. infundibuliformis
Cap 5–10cm/2–4in. The
cap is a delicate pinkish-
buff to tan and smooth,
while the decurrent gills
are whitish, as is the stem.
This rather delicate, thin-
fleshed mushroom is very
common in mixed woods,
especially deciduous,
throughout Britain.
Spores white, 5–8 x
3.5–5μm. Edible but not
recommended.

Clitocybe nebularis

(Batsch) Kummer
Cap 5–15cm/2–6in. The
overall smoky-brown colours,
with crowded but hardly
decurrent gills, are distinctive
features along with the late
occurrence in the season in
piles of leaf-litter. **Spores**
pale buff, 6–7 x 3–4μm. It
often grows in fairy rings and
is common throughout
Britain from October to
December. Edible to many
but can cause upsets to
others, best avoided.

Clitocybe odora

(Bull. ex Fr.) Kummer
Cap 2.5–8cm/1–3in. The
delicate blue-green to grey-
green colours, lovely odour of
anise, smooth cap and pinkish-
buff spores all characterize this
beautiful mushroom. **Spores**
6–7 x 3–4μm. Common in
mixed woods throughout
Britain. Edible.

Clitocybe rivulosa

(Pers. ex Fr.) Kummer
Cap 2.5–5cm/1–2in. The whitish cap has a 'frosty' coating which is usually cracked in concentric zones showing a browner undersurface. The crowded gills are white and adnate. **Spores** white, 4–5.5 x 2.5–3μm. A frequent species in grassy areas late in the season and widely distributed. An extremely poisonous species to be avoided, as are all small, white mushrooms.

Clitocybe nuda

(Bull. ex Fr.) Big. & A. H. Smith
Cap 5–15cm/2–6in. The lovely smooth, violet cap soon fades to a violet-tan but the crowded gills remain a pale lavender-violet. The stem is tough, fibrillose and pale violet. The odour is often rather fragrant. **Spores** pale pinkish-buff, 6–8 x 4–5μm and minutely roughened. A common species, often in large numbers in circles on beds of leaf–litter or compost throughout Britain. Edible and very popular.

Clitocybe inversa

(Scop. ex Fr.)
Cap 5–10cm/2–4in. The funnel-shaped cap is a rich tawny-orange to foxy-red. The decurrent gills are a paler orange. **Spores** pale creamy-yellow, 4–5 x 3–4μm, minutely roughened. Often in groups, especially under conifers, quite common. Edible but not recommended.

Laccaria laccata
(Scop. ex Fr.) Bk. & Br.
Cap 1–5cm/½–2in. One of the most
confusing mushrooms because it has
many variations in colour, size and
shape. However, if the characters of
thick, pinkish gills, pinkish-brown to
reddish-brown cap and fibrous stem,
and white spore-print are observed
then identification is not too difficult.
Spores about 8µm, globose, with
spines up to 1 µm. One of the most
common and widespread species over
the northern hemisphere, found in
mixed woods, bogs and open
moorland with trees. Edible.

Laccaria bicolor
(R. Maire) Orton
Cap 2.5–5cm/1–2in. The pale
ochre-buff cap is smooth to
slightly roughened. Gills are
thick, widely spaced and a pale
pinkish-lilac. The fibrous stem is
coloured like the cap except for
base which is usually a bright
violet-lilac. **Spores** white, 7–9 x
6.5–7µm, with sharp spines up to
1µm high. Locally common in
mixed woods. Edible.

Laccaria amethystea
(Bolt. ex Hooker) Murrill
Cap 2.5–5cm/1–2in. The
beautiful deep amethyst-violet
of the entire fungus when fresh
and moist is quite unmistakable,
but as it dries the cap becomes
a dull, greyish-lavender to
almost white. The gills however
remain deep violet. **Spores**
white, globose, 8–10µm, minutely
spiny. Common in shady, damp
woodlands throughout Britain.
Edible.

Laccaria proxima

(Boud.) Pat.
Cap 2.5–8cm/1–3in. The reddish-
ochre to orange-brown cap is finely
scaly and becomes paler as it drys out.
The thick gills are pale pink. The tall,
fibrous stem is coloured like the cap.
Spores 7.5–10 x 6.5–7.5µm, with
spines up to 1.5µm. A common
species often confused with *L. laccata*
but distinguished by its large size and
elliptical spores. Usually found in
open woods or moorlands, often in
sphagnum moss. Edible.

Laccaria tortilis

(Bolt.) S. F. Gray
Cap 0.6–1cm/¼–½in. Perhaps the
smallest species of *Laccaria*,
distinguished by the thick gills which
are few in number, and the wavy-
fluted cap which is pale pinkish-
brown. **Spores** white, globose,
11–16µm and spiny, and with only 2
spores per basidium. Rather
uncommon, in very damp areas, often
on bare soil in deep shade. Edible but
worthless.

Tricholoma sulphureum

(Bull. ex Fr.) Kummer
Cap 5–10cm/2–4in. The
overall sulphur-yellow
coloration is combined
with a strong and
pungent odour of coal-
gas, making an easily
recognized combination.
Spores white, 9–12 x
5–6µm. Quite common,
usually under oaks or
beech, on acid soils.
Inedible, possibly
poisonous.

Tricholoma saponaceum
(Fr.) Kummer
Cap 5–10cm/2–4in. This is
a very variable mushroom
with colours ranging from
grey-brown to greenish-
brown and often with pink
flushes in the stem base. The
cap is smooth to minutely
scaly. The odour is rather
fragrant, soapy. **Spores**
white, 5–6 x 3.5–4µm. A
common species throughout
Britain, under mixed trees.
Inedible, possibly poisonous.

Tricholoma sulphurescens
Bresadola
Cap 5–15cm/2–6in. A large,
attractive species with white
cap and stem but with all
parts staining sulphur-yellow
when handled. The odour
can be rank and unpleasant
but the taste is mild. **Spores**
white, 5–6 x 4–5µm. This is
a rare species and its
distribution is uncertain, it is
found under oaks and
conifers. Inedible.

Tricholoma terreum
(Sch.) Kummer
Cap 5–10cm/2–4in. The
cap is dull grey to
brownish-grey, felty-scaly
with darker fibres. The gills
and stem are white. The
odour and taste are mild.
Spores white, 5–7 x
4–5µm. An often abundant
species under pines late in
the season, throughout
Britain. Edibility uncertain,
best avoided.

Tricholoma virgatum
(Fr.) Kummer
Cap 5–10cm/2–4in. The conical
cap is silvery-grey with darker
grey-black fibres and feels smooth
and dry. The gills are greyish to
yellow-grey and are notched where
they meet the stem. The firm stem
is fibrous, white. The taste of th
flesh is sharp and peppery. The
taste is sharp, and peppery.
Spores white, 6–7 x 5–6μm.
Often found in small troops under
conifers throughout Britain.
Usually considered inedible.

Tricholoma columbetta
(Fries) Kummer
Cap 5–10cm/2–4in. The snow-
white cap is quite smooth and silky
and slightly umbonate at the centre.
It becomes slightly buff when old
and may have small bluish or pink
spots in some locations. The gills
are white, and sinuate. The firm,
fibrous stem is also snow-white.
Spores white, 5–7 x 3.5–4.5μm. A
frequent under mixed trees,
especially in the south. Inedible.

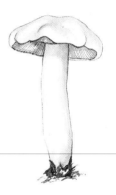

Tricholoma sejunctum
(Sow. ex Fr.) Quélet.
Cap 5–10cm/2–4in. The pale
greenish-yellow cap is marked
with darker, often blackish fibres.
The broad, widely spaced gilles are
white or flushed yellow and usually
deeply notched by the stem. The
fleshy, fibrous stem is white and may
be flushed yellow. Both the taste and
odour are mealy, cucumber-like, with
the taste becoming bitter with age.
Spores white, 5–6 x 4–5μm. A
frequent species under mixed trees,
especially in the south. Inedible.

Tricholoma flavovirens

(Pers. ex Fr.) Lund. (= *T. equestre*)
Cap 5–10cm/2–4in. The
yellow cap is marked with
dark brownish to olive fibres
and is usually dry. The stem
and gills are both yellow.
Spores white, 6–8 x 3–5μm.
Usually found under pines in
sandy soils, it is widely
distributed throughout Great
Britain, especially Scotland.
Edible and good.

Tricholoma portentosum

(Fr.) Quel.
Cap 5–12.5cm/2–5in. The
smooth, greyish cap is often
flushed with yellow and is
marked with darker, blackish
fibres. The stout, white stem
and white gills are often also
flushed with yellow. The odour
and taste are mealy, cucumber-
like. **Spores** white, 5–6 x
3.5–5μm. Found under
conifers, occasionally under
beech, most commonly in
Scotland. Edible.

Tricholoma fulvum

(D. C.) Sacc. (= *T. flavobrunneum*)
Cap 5–10cm/2–4in. The moist,
orange-brown to red-brown cap
often has a grooved margin.
The gills are yellowish spotted
with reddish-brown while the
stem is fibrous, reddish-brown.
The flesh in the stem is yellow.
Easily distinguished by the
yellow flesh and speckled gills.
Spores white, 5–7 x 3–4.5μm.
A common species under birch
in wet areas, throughout
Britain. Inedible.

Tricholoma aurantium
(Sch.) Ricken

Cap 5–10cm/2–4in. This attractive species is a bright orange to orange-brown with a viscid cap with darker brownish flecks. The gills are white, often spotted orange, while the stem has concentric orange-scaly zones over the lower half. Odour and taste are mealy cucumber-like. **Spores** 4–5 x 3–3.5µm. Found mainly in the Highlands under pine. Inedible.

Tricholoma caligatum
(Viv.) Ricken (= *Armillaria caligata*)

Cap 5–15cm/2–6in. The dull, pallid cap has darker brown scales and fibres. The stem is white with a sheath of brown veil broken up into patches and zones over the lower half. There may be bluish stains present in some forms. **Spores** white, 6–7.5 x 4.5–5.5µm. The odour is often spicy or pungent while the taste can be disagreeable. Found mainly in Scotland under pines, often in large numbers. Edible.

Leucopaxillus giganteus
(Fr.) Singer

Cap 10–40cm/4–16in. This often huge mushroom is white to pale buffy-cream. The cap is soon rather funnel-shaped and is smooth to finely suede-like. The margin is inrolled and grooved or furrowed. The gills are decurrent down the short, stout stem. **Spores** white, 6–8 x 4–5.5µm. Often found growing in large circles in meadows and grassy woodlands, it is rather uncommon. Edible.

Melanoleuca melaleuca
(Pers. ex Fr.) Maire
Cap 5–10cm/2–4in. The
smoky brown to dark
brown cap is usually
slightly domed or umbonate
at the centre and is smooth. The
gills are sinuate, white to cream,
crowded. The stem is slightly paler
than the cap, fibrous and with the
flesh in the base often a darker
brown. **Spores** white, 7–8.5 x
5–5.5μm, amyloid and minutely
warted. A common species
rather late in the season in mixed
woodlands and open spaces everywhere.
Edibility uncertain, best avoided.

Melanoleuca cognata
(Fr.) Konrad & Maubling
Cap 5–10cm/2–4in. The
domed cap is a rich yellow-
ochre to yellow-brown while the gills
are deep ochre. The fibrous stem is
paler, cream-ochre. The odour can be
floury-rancid. **Spores** are unusual for
the group, being deep creamy-yellow,
9–10 x 6–7μm, with amyloid warts.
Quite common in open woodlands,
widely distributed. Edible.

Lyophyllum decastes
(Fr.) Singer (= *L. aggregatum*)
Cap 5–10cm/2–4in. Often growing
in enormous clumps the dull grey-
brown caps are smooth and
sometimes irregular in shape. The
gills are white to buff and attached
to the stem, sometimes with a
decurrent tooth. The tough,
fibrous stems are white or grey.
Spores white, globose, 4–6μm.
Usually found in disturbed soils on
roadsides, gardens, woodland
edges, throughout Britain. Edible

Calocybe gambosa St. George's Mushroom
(Fr.) Donk = *Tricholoma gambosum,*
T. georgii

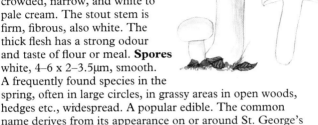

Cap 5–15cm/2–6in. The thick,
fleshy cap is white to pale ivory
or tan, smooth with an incurved
margin at first. The gills are
crowded, narrow, and white to
pale cream. The stout stem is
firm, fibrous, also white. The
thick flesh has a strong odour
and taste of flour or meal. **Spores**
white, 4–6 x 2–3.5μm, smooth.
A frequently found species in the
spring, often in large circles, in grassy areas in open woods,
hedges etc., widespread. A popular edible. The common
name derives from its appearance on or around St. George's
Day – April 23rd.

Calocybe carnea
(Bull. ex Fr.) Kuhn.

Cap 1–5cm/½–2in. This small, pink
to reddish–brick mushroom has
smooth, dry cap and stem. The gills
are white, crowded, attached to
slightly descending. The stem is
fibrous in texture. **Spores** white, 4–6
x 2–3μm, smooth. This pretty species
grows in grass in lawns and open
woodland clearings, rather uncommonly,
throughout Britain. Inedible.

Tricholomopsis rutilans
(Schff. ex Fr.) Singer

Cap 5–15cm/2–6in. The
contrasting colours of bright wine-
red cap and stem, against the
yellow gills, and the habitat on
conifer wood, allows easy
recognition. The cap is minutely
velvety-scaly. **Spores** white, 7–8 x
5–6μm. A common species on
dead and decaying pine stumps
and logs throughout Britain.
Edible but poor.

Tricholomopsis platyphylla
(Pers. ex Fr.) Singer

Cap 5–15cm/2–6in. One of the earliest mushrooms to appear, the dull grey-brown cap is streaked with radial fibres. The gills are very broad, widely spaced and often with split or jagged edges. The white stem is very tough and fibrous and has white root-like cords at the base (rhizomorphs). **Spores** white, 7–9 x 5–7μm. Common on deciduous logs, stumps and buried wood throughout Britain. Inedible.

Asterophora lycoperdoides
(Bull. ex Mer.) Dit.

Cap 0.6–2cm/¼–¾in. A remarkable small mushroom found on the decaying remains of old fungi, specifically *Russula* and *Lactarius*. The cap is covered in a thick brown powdery coating which is actually masses of asexual spores. The thick gills are widely spaced and often malformed. Widely distributed throughout Britain. Inedible. The related *A. parasitica* differs in forming a perfect small, white mushroom without the powdery coating, and also grows on rotting *Russula* and *Lactarius*.

Flammulina velutipes
(Fr.)Kar.

Cap 2.5–10cm/1–4in. Often fruiting while snow is still on the ground, the rich yellow-orange caps are slightly sticky. The gills are crowded, cream while the stem is yellow with a blackish-brown, hairy base. **Spores** white, 7–9 x 3–6μm. Growing in dense clusters on standing trees and stumps, especially elm, aspen and willow, throughout Britain, from October to May. Edible and good.

Cystoderma terrei

(Bk. & Br.) Harmaja = *C. cinnabarinum*
Cap 2.5–8cm/1–3in. The rich
brick-red to orange cap is very
powdery-granular as is the stem.
The gills and stem apex are
white. **Spores** white, 3.5–5 x
2.5–3μm. Uncommon in mixed
woods throughout Britain.
Inedible.

Cystoderma amianthinum

(Scop.) Fayod
Cap 2.5–5cm/1–2in. The
powdery-granular cap is bright
yellow-ochre and usually
wrinkled at the centre. The
stem has a yellow, granular
coating up to a slight ring-zone
near the apex. The gills are
white to cream. **Spores** white,
4–6 x 3–4μm. Common in
conifer woods throughout
Britain. Inedible.

Xerula radicata

(Relh.) Redhead
= *Oudemansiella radicata*
Cap 5–12.5cm/2–5in. This
tall, elegant species, if
carefully dug out of the soil,
will reveal a long, deeply
rooting 'tap–root'. The pale
brownish cap is glutinous in
wet weather but minutely
hairy when dry. The stem is
dry and minutely velvety.
Spores white, 12–18 x
9–12μm. Common around
dead stumps throughout
Britain. Inedible.

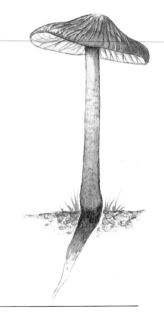

Oudemansiella mucida
(Schrad.) V. Hohn

Cap 5–10cm/2–4in. This very slimy mushroom is at first a greyish-brown when very young but soon becomes paler until the cap is a pure, shining white. The widely spaced gills are deep and also white. The tough stem is white to pale brown below and roots into the dead wood on which it grows. There is a distinct ring on the upper part. **Spores** white, 14–18 x 12–16μm, smooth. This beautiful mushroom grows in groups on dying or fallen beech trees in the late autumn and is one of the most striking fungi to be found. Inedible.

Marasmius androsaceus
The Horsehair Fungus
(L. ex Fr.) Fr.

Cap 0.6–1cm/¼–½in. The small, fluted brown cap on the thin, hair-like, blackish-brown stem are distinctive as are the very widely spaced gills. There is no odour, unlike some other *Marasmius* species. **Spores** white, 6–9 x 2.5–4.5μm. Often abundant on conifer needles and twigs in Britain. Inedible.

Marasmius rotula
(Scop. ex Fr.)Fr.

Cap 0.15–1cm/¹⁄₁₆–½in. The tiny, bell-shaped cap is white, pleated and sunken at the centre. The white gills are widely spaced, attached to the stem or separated by a narrow 'collar'. The thread-like stem is dark brown and shiny. It has no odour. **Spores** white, 6–10 x 3–5μm. Common on dead wood and twigs of deciduous trees throughout Britain. Inedible.

Marasmius oreades The Fairy Ring Mushroom
(Bolton) Fr.

Cap 2.5–8cm/1–3in. Although
other mushrooms also grow in rings
this is the one most people are aware
of since it disfigures lawns and grassy
areas throughout much of the world.
The tough, pale buff caps and fibrous
stems appear whenever the weather is
mild and damp. The gills are thick
and widely spaced. **Spores** white,
7–10 x 4–6µm. Edible and delicious
although care should be taken not to
confuse it with poisonous *Clitocybe*

species which also grow in lawns. The latter are usually white,
flattened to depressed caps with crowded gills.

Marasmius alliaceus
(Jacq. ex Fr.) Fries

Cap 2.5–5cm/1–2in. The thin, dull
ochre to whitish cap is dry and matt
or slightly wrinkled. The thin gills are
distant and whitish. The often tall,
thin stem is blackish-brown and
minutely velvety. The flesh when
crushed has a strong odour of garlic.
Spores pale cream, 7–11 x 6–8µm,
smooth. Quite common on fallen
beech leaves, especially in southern
England. Inedible.

Micromphale foetidum
(Sow. ex Fr.) Singer

Cap 1–2.5cm/½–1in. This small,
dark brown to red-brown species has
a flattened, wrinkled cap. The distant
gills are pale brown. The short,
slender stem is blackish-brown and
velvety. The whole mushroom has a
strong, unpleasant odour rather like
rotting cabbage or garlic. **Spores**
white, 8.5–10 x 3.5–4µm. A rather
uncommon species found on fallen
twigs and branches, widely
distributed. Inedible.

Collybia maculata
(A. & S. ex F.) Kumm.
Cap 5–10cm/2–4in. This all-white mushroom is very tough and fibrous, especially the stem which may be rather rooting in the soil. As it ages the cap develops spots and streaks of rust-brown. The taste is rather bitter.
Spores pale pinkish-buff, 5–6 x 4–5μm. Often in groups in the leaf-litter of mixed woods, widespread. Edible but tough.

Collybia butyracea
(Bull. ex Fr.) Quélet
Cap 5–8cm/2–3in. The domed cap is smooth and very greasy-waxy to the touch, it also becomes paler as it loses moisture and is often two-toned. The tough stem is club-shaped and fibrous. The crowded gills are often slightly jagged on the edges. **Spores** cream-buff, 6–8 x 3–3.5μm. Common under pines and widely distributed. Edible.

Collybia dryophila
(Bull. ex Fr.) Kumm.
Cap 2.5–8cm/1–3in. One of the most common fungi everywhere the pale yellow-brown to reddish cap is smooth and dry. The crowded gills are whitish while the smooth stem is reddish-brown below with white or yellow hairs at the base. **Spores** pale cream, 5–6 x 2–3μm. Often abundant in mixed woods, especially under oaks and pines, throughout Britain. Edible.

Collybia fusipes
(Bull.) Quélet
Cap 5–10cm/2–4in. The
fleshy, smooth caps are a
dull brick-red to reddish-
brown. the thick, fleshy gills
are widely spaced and pale
buff often spotted rust-
brown. The very tough,
fibrous and fleshy stem is
usually swollen and spindle-
shaped, rooting into the
ground. **Spores** white, 4–6 x
3–4.5µm. Found in large
clumps at the base of oaks
and beeches, very common
in southern England, scarcer
further north. Inedible.

Collybia confluens
(Pers. ex Fr.) Kumm.
Cap 2.5–5cm/1–2in. The dull
grey-brown to pinkish caps and
pale brown stems, which are
densely hairy and grow in large
clusters, are the principal
characters to look for. The gills are
narrow and very crowded. **Spores**
white, 7–10 x 2–4µm. On fallen
leaves or needles in mixed woods,
widely distributed. Edible but
poor.

Mycena galopus
(Pers.) Kummer
Cap 0.6–1cm/¼–½in. The small.
bell-shaped cap is greyish-black,
fading to whitish-grey. The gills are
pale grey while the stem is darker,
grey-brown below. When broken the
stem exudes a drop of white, milky
fluid. **Spores** white, 10–14 x 5–7µm.
A common species in mixed woods,
especially coniferous, widely
distributed. Inedible.

Mycena galericulata
(Scop. ex Fr.) S. F. Gray
Cap 2.5–5cm/1–2in. The broadly
bell-shaped cap is greyish-brown to
buffy-brown and often radially
wrinkled. The gills are broad, white
flushed greyish-pink and often with
cross-veins between. The smooth
stem is grey also, and often deeply
rooting. **Spores** white, 8–11 x
5.5–7μm. This often very common
species grows in tufts on rotting
deciduous logs and stumps
throughout Britain Edible but not
recommended.

Mycena inclinata
(Fr.) Quélet
Cap 2.5–5cm/1–2in. Often
confused with *M. galericulata*
but the stem shades from
reddish-brown below to
yellowish above and has
minute whitish flecks below
(use a hand lens for this). The
cap margin is often minutely
toothed. The odour is distinct,
fragrant, soapy to slightly
rancid. **Spores** white, 8–10 x
5.5–7μm. Often very common
on deciduous tree stumps
throughout Britain. Edibility
doubtful.

Mycena haematopoda
(Pers. ex Fr.) Kummer
Cap 1–5cm/½–2in. The dark reddish-
brown to wine-coloured cap and stem
are distinctive as is the deep reddish-
brown juice which 'bleeds' from any
broken stem. **Spores** white, 9–10 x
6.5–7μm. A common species, growing
in clumps on deciduous wood
throughout Britain. Edibility
doubtful.

Mycena pura

(Pers. ex Fr.) Kummer
Cap 2.5–5cm/1–2in. Cap,
gills and stem range from a
delicate lilac or pale violet to
pinkish or even blue-grey.
What remains constant is the
strong odour and taste of
radish. **Spores** white, 5–9 x
3–4µm. A common species,
usually growing singly on leaf-
litter, throughout Britain.
Inedible; reported to cause
poisoning.

Mycena polygramma

(Bull.) S. F. Gray
Cap 2.5–5cm/1–2in. The grey
to blue-grey or grey-brown
cap is furrowed at the margin.
The narrow gills are pale
greyish-white. The slender
stem is silvery-grey with fine
silky lines running lengthwise.
Spores white, 8–10 x
5.5–7µm. Found in small tufts
on fallen deciduous wood it is
a common species.
Inedible.

Mycena pelianthina

(Fr.) Quélet
Cap 2.5–5cm/1–2in. The
pinkish-lilac to grey-violet cap is
lined at the margin and soon
fades at the centre. The gills are
whitish to pale lilac-grey with a
distinctly darker, violet-brown
edge. The rather thick stem is
coloured like the cap. The whole
mushroom has a strong odour of
radish, especially when rubbed.
Spores white, 4.5–6 x 2.5–3µm.
Quite common in fallen leaves of
beech trees. Inedible.

Mycena sanguinolenta

(A. & S. ex Fr.) Kummer
Cap 0.6–1cm/¼–½in.
Another species which
'bleeds' reddish juice when
broken, but this species is
usually single, not clustered,
and always very small, unlike
M. haematopoda, also shown
here. The colours are bright
reddish-orange to reddish-
brown. **Spores** white, 8–11
x 4–6μm. Common on
decaying leaf or pine needle
litter, throughout Britain.
Inedible.

Mycena epipterygia

(Scop.) S. F. Gray
Cap 0.6–1cm/¼–½in. The
bell-shaped cap is sticky
and a pale greenish-yellow
to grey-brown. The gills
are white to pale lemon.
The slender stem is sticky
and a pale to bright lemon-
yellow. **Spores** white,
8–12 x 4–6μm. Common
in mossy areas under
conifers and under
bracken on moors and
heaths. Inedible. The
slightly darker, greener,
and stouter *M. epipterygiodes*
is found on dead conifer
stumps.

AMANITA FAMILY

(Amanitaceae)

There are over 20 species of *Amanita* in Britain and the genus contains some of the most beautiful, and the most deadly, mushrooms in the world. Features to look for are the white spores, the universal veil which may be left as a sac-like volva at the stem base or as mere warty remnants on the cap and stem base. Also a partial veil may be present as a skirt-like ring or annulus at the top of the stem. Always make sure to collect the entire stem base from under the ground so any volva can be seen. Many *Amanitas* have distinctive odours. The gills are free from the stem as shown in the cross-sections below.

Amanita phalloides Death Cap
(Fr.) Sacc.

Cap 5–15cm/2–6in. This deadly species causes deaths almost every year in Europe and also in North America where it has been introduced and is spreading rapidly. The olive-green to yellowish-green cap is smooth and has very fine radial streaks or fibres. Very rarely the cap may be brownish or even white. The gills are white, broad and free of the stem. The stem is bulbous and encased at the base in a large white, bag-like volva. At the top of the stem hangs a membranous ring. The flesh is white and as it ages develops a strong, rather sickly-sweet odour rather like old honey. **Spores** white, 8–10.5 x 7–8μm, amyloid. Found from late summer onwards especially in the south under oaks and occasionally conifers. Deadly poisonous, look out for tell-tale volva and free, white gills.

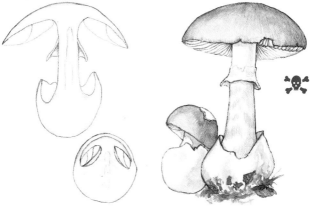

Amanita virosa Destroying Angel
Secr.

Cap 5–15cm/2–6in. The snow-white cap is often bluntly rounded to almost conical and slightly sticky-greasy. The pure white gills are free and quite crowded. The stem is white and has a slightly woolly-scaly surface with a large white volval sack at the base. A fragile, often torn ring is at the stem apex. The odour is faint to a little unpleasant, sickly-sweet of honey. KOH (caustic soda) placed on the cap turns bright yellow. **Spores** white, 9–11 x 7–9μm, amyloid. This rare species is found mainly in the Highlands especially under birch and is deadly poisonous!

Amanita submembranacea
Bon

Cap 8–10cm/3–4in. The grey-brown to olive-brown cap can be quite dark and has long striations at the margin; there are sometimes flakes of the universal veil clinging to the cap surface. The gills are broad, free and white. The stem is tall and faintly banded with yellowish-grey or grey-brown and has a thick, greyish-white volva at the base. **Spores** white, 9–12μm, rounded and non-amyloid. A rather uncommon although probably often misidentified species, most often found in Scotland. Inedible.

Amanita vaginata
(Bull. ex Fr.) Vitt.
Cap 5–10cm/2–4in. This
elegant species is one of the
'ringless' *Amanita* species,
sometimes called
Amanitopsis. The cap is a
delicate pale grey to steel-
grey with the margin deeply
radially grooved. The gills
are free of the stem, white to
very pale grey, their margins
often minutely ragged. The
tall stem is white with faint
grey bands and has a white
volval sac at the base.
Spores white, 9–12um,
globose, non-amyloid.
Frequent in mixed woods
throughout Britain. Edible
but best avoided.

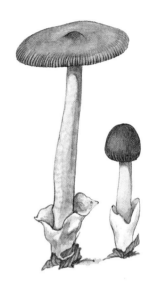

Amanita fulva Tawny Grisette
(Schaef. ex Fr.) Pers.
Cap 5–10cm/2–4in. The rich
tawny to reddish-brown cap or
orange-brown cap has a central
umbo and deep marginal
grooves. The gills are cream,
quite crowded, and with the
margins often minutely ragged.
The tall stem is cream sometimes
with darker bands and emerges
from a thick white volval sac. The
inner surface of the sac may be
coloured pale brown. There is no
ring present on the stem.and
long, elegant stem emerge from a
white volval sac. There is no ring
on the stem. **Spores** white,
8–10um, globose, non-amyloid.
Common or even abundant in
both deciduous and coniferous
woodlands throughout Britain.
Edible but best avoided like all
Amanita species..

Amanita ceciliae

(B. & Br.) Bas = *A. inaurata*
Cap 5–12.5cm/2–5in. The
dark brown cap usually has
many darker fragments of veil
left on the surface. The gills
are white to greyish and the
tall greyish stem has a grey,
sac-like veil which soon
breaks into small fragments,
often left behind in the soil.
Spores white, 11.5–14μm,
globose, non-amyloid. A rare
species in mixed woods
throughout Britain. Edibility
uncertain.

Amanita battarae

(Boudier) Bon = *A. umbrinulutea*
Cap 8–12.5cm/3–5in. This
ringless species is quite
large with an olive-brown
to yellow-brown cap with a
distinctly grooved margin
and usually a darker
coloured zone behind the
grooves. The gills are
cream, broad and free of
the stem. The stem is often
banded with slightly darker
yellow-brown on a pale
buff background. There is
a prominent, thick white to
ochre volva at the base.
Spores white, 11–16 x
9.5–13μm, non-amyloid.
This rather rare species in
Britain is found in
coniferous woods,
particularly in the south.
Inedible.

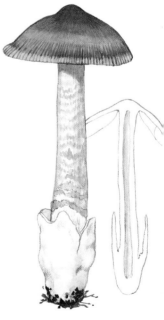

Amanita crocea
(Quélet) Singer

Cap 8–12.5cm/3–5in. This species is a beautiful pale orange to orange-brown with a striate cap margin. The gills are cream to very pale orange, free. The tall stem is slightly paler than the cap and often banded with darker orange scales, there is a thick white volva at the base, usually with the inner surface a pale orange-buff. **Spores** white, 8–12µm, rounded, non-amyloid. A rather uncommon but lovely species it appears most frequent in Scotland although large fruitings have also been seen in the south of England. Inedible.

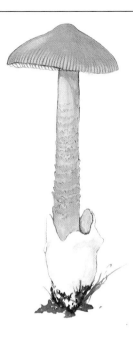

Amanita citrina var. alba
(Gillet) Gilbert

Cap 8–12.5cm/3–5in. The pure white cap is smooth but usually has many remnants of the enclosing universal veil scattered over the surface. The gills are white, broad and free. The stem is white, very swollen and rounded, bulbous at the base with a distinct, gutter-like margin. A well-developed ring hangs down at the top. The flesh has a strong odour of freshly dug potatoes. **Spores** white, 7–10µm, rounded, amyloid. Found (exclusively) under beech, widespread. Inedible. Although usually considered a variety of *A. citrina* (see below) it may be best considered as a species in its own right. It is consistently larger and more robust and very restricted in its habitat; *A.citrina* grows under a wider range of trees including conifers.

Amanita citrina
Schaeff. ex S.F. Gray

Cap 5–10cm/2–4in. The pale lemon-yellow cap has numerous patchy remains of the universal veil, usually whitish-ochre in colour. The gills are white and free. The stem may also be tinted pale yellow with a large, distinctly margined bulb at the base and a pale yellow ring at the apex. The flesh smells strongly of freshly dug potatoes. **Spores** white, 7–10μm, rounded, amyloid. Under mixed trees, beech, oak, birch, even conifers. Inedible.

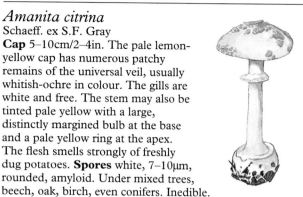

Amanita porphyria
(A. & S. ex Fr.) Secr.

Cap 2.5–8cm/1–3in. The greyish-brown cap has faint purplish tints although these are more obvious on the stem and the ring. The stem ends in a large, rounded bulb which has a margin on the upper edge. The gills are white and free. The odour may be of potato or radish. **Spores** white, 7–9μm, globose. A fairly uncommon species found in mixed woods throughout Britain. Possibly poisonous.

Amanita eliae
Quélet

Cap 5–10cm/2–4in. The domed cap is a pale pinkish-ochre with short, fine lines at the margin. The gills are white. The stem is white, usually with the long, rooting base sunk quite deeply into the soil. The stem does not have a volva, but there is a fragile ring at the top. **Spores** white, 11–14 x 6.5–8.5μm, non-amyloid. This rare species grows in deciduous woods on acid soils. Inedible.

Amanita muscaria Fly Agaric
(L. ex Fr.)
Cap 8–25cm/3–10in. This bright red
mushroom is probably the most well-
known fungus in the world, appearing in
numerous children's books and fairy tales.
The white or yellowish spots on the cap are
the remains of the universal veil. The stem
and gills are white, while the stem has rings
of white or yellow warts at the swollen base.
A prominent, floppy ring is present. **Spores**
white, 9.5–13 x 6.5–8.5μm, non-amyloid.
This often very common species is found
throughout Britain especially under birch
and sometimes under pines. Poisonous
although rarely deadly, it was traditionally
used to poison houseflies, hence the common name.

Amanita pantherina
(DC. ex Fr.) Secr.

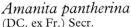

Cap 5–10cm/2–4in. The brownish cap has pure
white fragments of veil and the gills and stem are
also white. The abruptly bulbous stem has a narrow
floppy ring above and one or more rings of veil tissue
just above the veil. **Spores** white, 8–14 x 6.5–10μm,
non-amyloid. Uncommon, found in mixed woods
throughout Britain. Very poisonous, causing
delirium and coma-like deep sleep.

Amanita friabilis
Karsten

Cap 2.5–8cm/1–3in. This often very
small species is a pale grey-brown to
sepia with woolly tomentose veil
remnants on cap and stem. The gills
are white to pale greyish, free. The
stem is pale greyish, bulbous but
without a volva, sometimes with
fragile remnants of veil at the base.
Spores white, 9.5–14 x 7–10.5μm,
non-amyloid. This rare (but possibly
overlooked) species is only found in
alder woods, its distribution is
uncertain but possibly widespread.
Edibility unknown.

Amanita gemmata

(Fr.) Gill. = *A. junquillea*
Cap 5–10cm/2–4in. The buffy-
yellow cap has white patches of
veil and the gills and stem are
white. The base of the stem is
bulbous with an abrupt margin
or 'gutter' on the upper edge.
Spores white, 8.5–11 x
5.5–8.5μm, non-amyloid.
Frequent in oak and pine
woods, throughout Britain.
Possibly poisonous.

Amanita rubescens The Blusher

(Pers. ex Fr.) S. F. Gray
Cap 5–15cm/2–6in. The
common name refers to the
pinkish-red blush that occurs
wherever the fungus is bruised
or damaged, and with age
(splitting the stem usually
shows reddish worm holes).
The cap varies from yellowish-
brown to reddish-brown.
Spores white, 8–10 x 5–6μm,
amyloid. Very common in
mixed woods, widely
distributed in Britain. Edible
but best avoided.

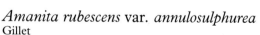

Amanita rubescens var. *annulosulphurea*

Gillet
Cap 5–8cm/2–3in. Usually
regarded as a variety of the
previous species but some
authors consider it as a
species in its own right. The
variety is usually smaller,
paler and has a bright
yellowish ring. **Spores** the
same. It is quite uncommon
and grows in mixed woods.
Apparently edible but to be
avoided.

Amanita spissa

Fr.) Kumm.

Cap 5–15cm/2–6in. The greyish cap has fine, irregular greyish remnants of veil. The gills are white while the stem is white with greyish overtones. The base of the stem is swollen and slightly rooting. The odour is often of radish. **Spores** white, 9–10 x 7–8μm, amyloid. Very common throughout Britain in mixed woods. Edibility uncertain, best avoided.

Amanita francheti

(Boud.) Fayod = *A. aspera*

Cap 5–15cm/2–6in. This essentially brown-capped mushroom is covered in bright yellow fragments of veil and these fragments are also found on the base of the stem and on the edge of the ring at the stem apex. **Spores** white, 8–10 x 6–7μm, amyloid. Uncommon, found in deciduous woods especially in the south. Edibility uncertain, best avoided.

Amanita strobiliformis

(Vitt.) Quélet = *A. solitaria*

Cap 10–25cm/4–10in. This sometimes huge species is completely white. The cap is covered in large, soft flattened warts or scales, discolouring slightly greyish with age. The margin has ragged fragments hanging down. The gills are white, crowded. The robust stem is cylindrical to slightly swollen and often deeply rooted, the surface of the stem is shaggy and at the top is a fragile, torn ring with the consistency of cream cheese. **Spores** white, 9–14 x 7–9.5μm, amyloid. This striking species grows under beech trees on chalky soils and is not uncommon in parts of southern England. Apparently edible but to be avoided.

Amanita echinocephala
(Vitt.) Quélet

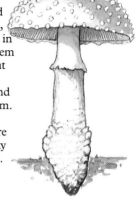

Cap 8–12.5cm/3–5in. The rounded
cap is white and covered with small,
conical warts. The gills are unusual in
having a pale greenish flush. The stem
is bulbous and has bands of scales at
the base and a thin ring at the top.
The flesh has an unpleasant taste and
odour, rather alkaline or like old ham.
Spores white to slightly greenish,
9.5–11 x 6.5–7.5μm, amyloid. A rare
species found under beech on chalky
soils especially in southern England.
Inedible.

Chamaemyces fracidus
(Fr.) Donk

Cap 2.5–8cm/1–3in. The ivory-
white to buff cap is slightly wrinkled
and usually marked with darker,
reddish-brown patches. The gills are cream,
crowded, free of the stem and usually weep
dark droplets of liquid. The stem is coloured
like the cap and has a rough, woolly surface
up to a faint ring zone at the top. Like the
gills the stem may weep droplets. The flesh
is cream with a faint rubbery smell. **Spores** pale buff, 4.5–5 x
2–2.5μm. This species is not uncommon in grass at woodland
edges especially in the south. Inedible.

Limacella glioderma
(Fr.) Maire

Cap 2.5–8cm/1–3in. The very slimy cap is
dark reddish-brown to pinkish at the margin,
the centre is often granular under the slime. The
gills are white and free while the stem is light
brown with a very slimy veil over the surface. The
odour is strong mealy. **Spores** white, 3–4μm,
globose, non-amyloid. A rather rare species,
found under hemlock and birch, widespread.
Edibility unknown, best avoided. A close
relative of the genus *Amanita* differing
principally in the glutinous veil, although some
mycologists think it is closer to the following genus, *Lepiota*.

LEPIOTA FAMILY

(Lepiotaceae)

These mushrooms have gills free from the stem, white to pale pinkish-cream, or green spores and usually a more or less obvious ring on the stem. They are closely related to the black-brown spored *Agaricus* mushrooms.

Lepiota ignivolvata

Bousset-Joss.

Cap 8–12.5cm/3–5in. The whitish-buff cap has a darker brown central boss, the surface is finely scaly. The gills are white and free of the stem. The stem is club-shaped with one or two oblique brownish rings, the base flushes slowly reddish. The flesh is white with a rubbery odour. **Spores** white, 11–13 x 5–6μm, spindle shaped. This uncommon species grows under conifers on calcareous soils. Edibility unknown

Lepiota felina

(Pers.) Karsten

Cap 2.5–4cm/1–½in. The delicate white cap has dark, blackish-brown scales spreading out from the centre. The thin, crowded gills are white and free. The slender stem is white with small blackish scales below the thin, black edged ring. The flesh has an odour of cedarwood. **Spores** white, 6.5–7.5 x 3.5–4μm, oval. Uncommon but found in coniferous woods, widely distributed. Edibility unknown, possibly poisonous.

Lepiota ventriosospora

Reid

Cap 5–8cm/2–3in. The ochraceous cap has darker yellow-brown scales and is often bright yellowish at the margin. The gills are white, free. The stem is slender, coloured like the cap and has a woolly-scaly coating. **Spores** white, 14–18 x 4–6μm, spindle shaped. This uncommon species can grow under both conifers and deciduous trees and is widespread in Britain. Edibility unknown.

Lepiota castanea
Quélet

Cap 2.5–5cm/1–2in. Chestnut brown
to rust brown with closely packed
scales. The gills are white sometimes
stained rust. The slender stem has
small rust scales over the lower half
up to a thin ring zone. **Spores** white,
9–13 x 3.5–5μm, bullet-shaped with
the apiculus offset. Quite common in
mixed woods especially in the south.
Inedible.

Lepiota brunneoincarnta
Chod. & Mart

Cap 2.5–5cm/1–2in. The rounded
cap is a dull pinkish-brown to
purplish-brown with darker purple-
brown scales. The gills are white. The
rather short, stout stem is whitish
flushed pinkish-purple below a faint
ring zone. There are usually one or
more darker scaly bands below the
ring. **Spores** white, 7–9 x 4–5μm,
oval. This uncommon species is found
in grassy clearings, woodlands and
sometimes in urban gardens and is
known to be very poisonous, even
deadly.

Lepiota cristata
Fr.) Kumm.

Cap 1–5cm/½–2in. The small, white
cap is marked with concentric rings of
dark, reddish-brown scales. The white
gills are free of the stem and the
slender white stem has a tiny white
ring towards the top. The odour is
strong, unpleasant, of rubber or of the
common Earthball *Scleroderma*.
Spores white, 5–7 x 3–4μm, wedge-
shaped, dextrinoid. Frequent in grassy
areas in woods and pathsides
throughout Britain. Possibly
poisonous.

Lepiota clypeolaria

(Bull. ex Fr.) Kumm.
Cap 2.5–8cm/1–3in. The cap
has small, brownish scales
while the yellowish-brown
stem is very shaggy-woolly up
to a slight ring-zone. The free
gills are white. **Spores** white,
12–20 x 4–5μm, spindle-
shaped, dextrinoid. In mixed
woods throughout Britain.
Poisonous.

Cystolepiota aspera

(Pers.) Bon = *Lepiota aspera*
Cap 5–15cm/2–6in. The bluntly
conical cap is a dark ochre-brown
with darker brown conical warts. The
crowded gills are white and often
forked, with rather fuzzy edges. The
stout stem is slightly bulbous and
white with darker brown scales at the
base. There is a floppy ring with dark
scales on the edge on the upper part
of the stem. The flesh has a strong
odour of rubber or of the common
Earthball *Scleroderma*. Quite common
under beech in England.
Inedible.

Cystolepiota bucknallii

(Bk. & Br.) Singer & Clemç.
= *Lepiota bucknallii*
Cap 1–2.5cm/½–1in. The
whitish-ochre cap has a flush of
lilac at the margin and has a
powdery surface. The gills are dull
ochre, widely spaced. The slender
stem is almost smooth, creamy
ochre with a lavender base. The
odour is strong of coal gas. **Spores**
white, 7–8 x 3μm. This species
may be locally common on rich
soils in deciduous woods
throughout Britain. Inedible.

Cystolepiota seminuda
(Lasch) Bon

Cap 1cm/½in. This small, delicate species is white with a powdery surface. The gills are white. The slender stem is white flushed lilac below. The flesh is white and odourless. **Spores** white, 3.5–4 x 2.5–3µm. Common on damp soil in woods and along roadsides throughout Britain. Inedible.

Melanophyllum haematospermum
(Bull.) Kreisel = *M. echinatum*
= *Lepiota echinata*

Cap 2.5–5cm/1–2in. The powdery cap is dark yellow-brown to grey-brown with a ragged margin. The gills are pinkish to dark wine-red. The stem greyish-ochre to brown, flushed pinkish-red with a granular surface. **Spores** at first greenish then drying to dull reddish, 5–6 x 3–3.5µm. An uncommon but often overlooked species found on soil along roadsides, in nettle beds and gardens, damp woodlands. Inedible.

Melanophyllum eyrei
(Massee) Singer

Cap 1–2.5cm/½–1in. The dull white to creamy-ochre cap has a powdery-granular surface. The gills are unusual in being a clear blue-green. The slender stem is coloured like the cap. **Spores** pale green, 3.5–4 x 2.5µm. Found on damp soil in deciduous woods, usually under low, shady herbage, uncommon. Inedible.

Leucocoprinus birnbaumi
(Corda) Singer
Cap 1–2.5cm/½–1in. This beautiful, fragile, all-yellow mushroom is found only in greenhouses or in pots of houseplants in Britain. **Spores** white, 8–13 x 5–8μm, elliptical, dextrinoid. This species probably originates from the tropics. Inedible.

Macrolepiota procera
(Scop.) Singer = *Lepiota procera*
Cap 12.5–25cm/5–10in. This very tall, stately species is one of our largest mushrooms. The brown, coarsely scaly cap, stem with brown banding, and thick, double-edged ring are features to look for. The gills are white and free. **Spores** white to cream, 15–20 x 10–13μm, elliptical. Found in open meadows, woodland edges and roadsides, especially in the south. Edible and delicious.

Macrolepiota excoriata
(Fries) Wasser
Cap 8–12.5cm/3–5in. The pale ochre cap has a felty, shaggy surface which breaks up and retracts at the margin to reveal the paler flesh below. The gills are white, free. The stem is white and smooth with a rather thin ring, two-layered ring above. **Spores** white, 12–15 x 8–9μm. Rather uncommon, it grows in open fields and pastures, especially in the south. Edible.

Macrolepiota gracilenta

(Fr.) Wasser = *Lepiota gracilenta*
Cap 8–10cm/3–4in. Slender,
graceful, this species has a
finely scaly, pale brown cap
and pale brown stem with
faint banding. The ring is thin,
often funnel-shaped when
young. The gills are white.
Spores cream, 10–13 x
7–8μm, elliptical. Quite
frequent in grassy areas in
woodlands, less commonly in
open fields. Edible and good.

Macrolepiota rhacodes

(Vitt.) Singer
Cap 8–15cm/3–6in. The shaggy brown cap, bulbous stem
which bruises deep reddish-brown, and the thick, double ring
are the distinctive features. **Spores** white, 6–10 x 5–7μm,
elliptical. In woods in deep leaf–litter, in conifer woods, or in
gardens on compost or heaps of leaves, common throughout
Britain. Edible and good.

Macrolepiota konradii

(Huijsman ex Orton) Moser
Cap 8–10cm/3–4in. The cap
is a dark grey-brown to
ochre-brown with the cap
cuticle breaking up at the
margin in a star-like manner
exposing the whitish cap
below. The broad gills are
white, free. The stem is
white, smooth, with very
fine, palest brown banding.
Spores cream, 11–15 x
8.5–9.5µm. Fairly common
in pastures and woodland
edges in southern England.
Edible and good.

Leucoagaricus leucothites

(Vitt.) Wasser = *Lepiota naucina*
Cap 5–10cm/2–4in. At first all-white,
this smooth-capped species eventually
ages to a dull greyish-white, The gills
bare white then greyish-pink. The
stem has a small, double-edged ring at
the top. **Spores** at first white, but
later deposits are pink, 7–9 x 5–6µm,
oval. In open fields, roadsides and
lawns, common throughout Britain.
Has caused some stomach upsets,
best avoided.

BLACK, DEEP BROWN TO PURPLE-BROWN SPORES

GOMPHIDIUS FAMILY

(Gomphidiaceae)

Although these mushrooms have gills they are related to the boletes with their spongy pores because of their very similar spores and other microscopic characters. Most species have a veil which may be dry or glutinous.

Gomphideus roseus

(Fr.) Karsten

Cap 2.5–5cm/1–2in. The beautiful coral-red to brick cap is smooth and very viscid, glutinous when wet. The thick, slightly decurrent gills are white then soon deep grey. The stem is white, tapering below, flushed slightly pinkish at the base and with a white, glutinous veil leaving a ring-zone at the top. The thick flesh is white, reddish below. **Spores** almost black, 15–17 x 5–5.5µm, long, spindle-shaped. Found under pines this lovely species is rather uncommon. Edible but poor.

Chroogomphus glutinosus

(Schaeff. ex Fr.) Fr.

Cap 5–10cm/2–4in. The pallid, grey-brown to purplish-brown cap is slimy, the decurrent gills are greyish-white and the stem is white with a yellow base and smeared with a slimy veil. **Spores** blackish, 15–21 x 4–7.5µm. Uncommon, found under conifers, especially spruce. Edible.

Chroogomphus rutilus
(Fries) Miller

Cap 5–10cm/2–4in. The orange-brown to copper-coloured cap often has a sharp umbo at the centre and is viscid when wet, drying smooth and shiny. The thick, decurrent gills are dull ochre then deep purplish-grey to black when mature. The stout stem is ochre-orange flushed pale wine-red above, with irregular woolly zones below a faint ring-zone. The flesh is ochre-orange. **Spores** almost black, 15–22 x 5.5–7μm, spindle-shaped. Frequent under pines, widely distributed. Edible.

AGARICUS FAMILY

(Agaricaceae)

Agaricus provides us with the well-known mushroom we purchase in stores, also the popular field mushroom. All species have deep brown spores, free gills, and a veil forming a ring on the stem.

Agaricus campestris Field mushrooms
L. ex Fr.

Cap 5–10cm/2–4in. The cap is white to greyish and may become fibrous-scaly with age. The gills are free and start a bright, rosy pink before turning deep brown as they mature. The short, stout stem is white with a fine, fragile ring. **Spores** deep brown, 6–9 x 4–6μm. A very common species, often growing in large fairy rings in open pastures and meadows. Edible and delicious.

Agaricus cupreobrunneus
(Moll.) Pilat
Cap 2.5–8cm/1–3in. The copper-brown cap is slightly fibrous-scaly. The gills start bright pink then mature deep brown. The short stem is white with a fragile, thin ring.
Spores brown, 7–9 x 4–6.5μm. In lawns, meadows, uncommon but widespread. Edible and good.

Agaricus bitorquis
(Quel.) Sacc. = *A. rodman*i
Cap 5–15cm/2–6in. The flattened white cap has extremely thick flesh compared with the very narrow gills. The latter start pinkish and turn deep brown. The stem is stout with a curious double ring on the lower stem, almost like a volva. **Spores** brown, 5–6 x 4–5μm. This species often pushes up through hard asphalt or gravel along roads, and prefers packed ground in urban areas. Edible and good.

Agaricus bernardii
(Quélet) Saccardo
Cap 12.5–20cm/5–8in. The thick, white cap is coarsely scaly and cracked and discolours greyish with age, the margin is inrolled. The crowded gills are very narrow, pale greyish-pink then dark brown. The short, thick stem is white with a thick,

sheathing ring. The flesh has a rather sour, fishy odour and stains reddish when cut. **Spores** dark brown, 5.5–7 x 5–6μm. An uncommon species found in meadows near the sea and also along roadsides (possibly where the roads have been salted in the winter). Edible but poor.

Agaricus augustus
Fries

Cap 10–25cm/4–10in. This
stately mushroom has a scaly-
fibrous, golden-brown cap. The
gills are whitish-pink then dark
brown. The tall stem has a large
floppy ring and is very woolly-
scaly below, when bruised it turns
dull yellow. The odour is pleasant
of anise. **Spores** brown, 8–11 x
5–6μm. Quite common in mixed
woodlands. Edible and excellent.

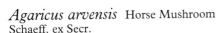

Agaricus arvensis Horse Mushroom
Schaeff. ex Secr.

Cap 10–15cm/4–6in. The cap is
smooth then slightly scaly with
age, the gills start white then turn
deep brown. The stout stem is
white and smooth and has a
large, floppy ring which has thick,
toothed or cog wheel-like tissue
on the underside. The odour is
pleasant of almond-anise. **Spores**
brown, 7–9 x 4.5–6μm. In fields
and woodland clearings, often
under spruce, throughout Britain.
Edible and delicious.

Agaricus essettei
Bow (= *A.abruptibulbus* in part)

Cap 8–15cm/3–6in. An all-white
species and rather elegant, the gills
start pale pink then mature brown.
The stem has a prominent, very
abrupt bulb. When bruised all
parts turn a dull yellow and the
flesh has a pleasant odour of anise.
Spores brown, 6–8 x 4–5μm.
Occasional in mixed woods,
especially spruce, throughout
Britain. Edible but must not be
confused with deadly white
Amanita species.

Agaricus silvaticus Red Staining Mushroom
Schff. & Secr.

Cap 5–10cm/2–4in. The scaly cap is
bright reddish- to yellow-brown
while the gills are pink then dark
brown. The stem is slender, bulbous
with a white ring, smooth on the
underside. The flesh when cut turns
more or less bright red. **Spores** brown,
4.5–6 x 3–3.5µm. In conifer woods ,
widespread. Edible. There are other red-
stainers such as *A. fuscofibrillosus* with fibrillose
(not scaly) reddish-brown cap and *A. benesi*,
an all-white species.

Agaricus placomyces
Peck

Cap 5–10cm/2–4in. This species is
rather slender with a finely fibrous-
scaly blackish-brown cap. The gills
are whitish-pink then brown. The
bulbous stem is white with a floppy
ring that often has yellowish droplets on
the underside. The odour is unpleasant,
like iodine. **Spores** brown, 5–7 x 3–4µm.
Uncommon, found in mixed woods and
along hedgerows, widespread. Not edible,
can cause serious upsets for some people.

Agaricus haemorrhoidarius
Kalchbr. & Schulz.

Cap 5–10cm/2–4in. The dull, dark
brown cap is densely fibrillose-scaly,
scales not very distinct. The gills are
crowded, bright pink at first then soon
dark brown. The stem is whitish to pale
brown, slightly felty on the surface, staining
bright red if scratched as will the cap also.
There is a white ring above which has brown
woolly flecks on the underside. The flesh is
white but stains bright red when cut and
bruised. **Spores** deep brown, 4.5–6 x 3–4µm.
Quite a common species in shady deciduous
woodlands especially under oak and beech.
Edible.

Agaricus xanthodermus Yellow Stainer
Gen.

Cap 8–12.5cm/3–5in. The white cap is often fibrous-scaly and turns dull greyish. The gills are whitish-pink then brown. The bulbous stem is white with a well-developed ring with cottony patches on the underside. The odour of the flesh when rubbed is strong, unpleasantly of iodine or old-fashioned school ink. When cut the flesh, especially at the base of the stem turns bright, chrome-yellow. **Spores** brown, 5–7 x 3–4μm. Frequent in urban areas by paths, hedges and in mixed woods, throughout Britain. Poisonous to many causing nausea, headaches.

Agaricus lutosus
(Møeller) Møeller

Cap 2.5–5cm/1–2in. This small species has a pale yellowish-brown cap which is finely scaly and very faintly flushed with purple on occasion. The gills are broad and pink before turning pale brown. The short, sturdy stem has a faint ring-zone. All parts bruise yellow on handling, especially the stem. **Spores** brown, 4.5–5.5 x 3.5–4μm. Found in short turf in lawns and fields, uncommon. Edible. This species appears to be identical with *A. micromegathus* Peck of North America and that name has precedence if this is proven.

STROPHARIA FAMILY

(Strophariaceae)

The gills are sinuate to adnate, purplish-brown, spores are deep purple-brown. The stem usually has a ring and the cap may be dry to glutinous.

Stropharia coronilla
(Bull. ex Fr.) Quélet

Cap 2.5–5cm/1–2in. The rounded cap is a pale yellow-ochre to bright yellow-orange, smooth and sticky when wet. The gills are sinuate, pale greyish when young then purple-brown when mature. The short stem is white, with a small ring often stained purple with spores. **Spores** deep purple-brown, 7–9 x 4–5μm. Quite common in open fields and meadows throughout Britain. Inedible.

Stropharia thrausta
(Schulz.) Sacc.

Cap 2.5–5cm/1–2in. An uncommon but beautiful species with bright reddish-orange colours on cap and stem, with small whitish scales present on the cap. **Spores** purplish-black, 12–14 x 6–7μm. Found on small twigs and fallen branches of deciduous wood, widely distributed. Inedible.

Stropharia semiglobata
(Batsch) Quélet

Cap 2.5–5cm/1–2in. The rounded, pale yellow cap is sticky when wet, while the tall, slender stem is only sticky below the faint ring-zone. The gills are greyish-lilac. **Spores** pale violet-brown, 15–19 x 8–10µm. Common on horse, sheep and cattle dung everywhere. Inedible.

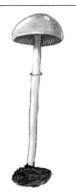

Stropharia aeruginosa
(Curt.) Quélet

Cap 2.5–8cm/1–3in. Truly green mushrooms are rare and this is one of the very few to be found. The sticky cap has prominent white flecks of veil and is yellower at the centre. The stem is also flecked-scaly with white on a green ground up to a ring-zone. The gills are deep violet-grey. **Spores** violet-brown, 7–9 x 4–5µm. On rich soil or in grass at the edge of woods, throughout Britain. Inedible. The similar and equally common *S. cyanea* differs in having very little white veil, and paler gills.

Psilocybe cyanescens
Wakefield

Cap 2.5–8cm/1–3in. The moist, almost greasy caps are very hygrophanous, which means they change colour markedly as they dry out. The rich reddish-brown colour fades to a dull ochre from the centre out. The gills are sinuate, pale buff then dark brown. The stem is white, fibrous and tough and stains pale blue on handling or with age. **Spores** dark purple-brown, 9–12 x 5–7µm. Often grows in enormous clumps on woodchip mulches in gardens and along woodland rides. This species was almost certainly introduced from America but is spreading here rapidly. Not edible, hallucinogenic and toxic.

Psilocybe semilanceata Liberty Caps
(Fr ex Secr.) Kumm.

Cap 0.6–2.5cm/¼–1in. The narrowly conical cap is smooth and sticky when wet, pale yellow-buff to brown. The narrow gills are greyish-brown and crowded. The stem is very slender and sinuous, whitish bruising blue when handled. **Spores** purple-brown, 11–14 x 7–8µm. In grass in fields and pastures, often abundant, throughout Britain. Toxic, hallucinogenic, the common name refers to the resemblance to French revolutionary hats.

Hypholoma fasciculare Sulphur Tuft
(Huds.ex Fr.) Kumm.

Cap 2.5–8cm/1–3in. The tufted growth, bright sulphur-yellow caps and stems, and the dark purple-brown spores are all distinctive features. The young gills are greenish-yellow before maturing to purple-brown. **Spores** 6.5–8 x 3.5–4µm. On logs, stumps and buried wood, throughout Britain, common. Poisonous.

Hypholoma sublateritium Brick Caps
(Fr.) Quélet

Cap 5–10cm/2–4in. These are also in tufts but they are larger, stouter and a deep brick-red. The gills are whitish then purple-grey. **Spores** purple-brown, 6–7 x 4–4.5µm. On dead stumps and logs, common in eastern Britain, late in the year. Edible but must not be confused with *H. fasciculare*.

COPRINUS FAMILY

(Coprinaceae)

Usually thin, delicate mushrooms (there are exceptions) they all have deep black or blackish-brown spores. Many species have a veil and/or a ring. *Coprinus* species are often called ink-caps because they dissolve away into an inky liquid as they disperse their spores.

Panaeolus sphinctrinus
(Fr.) Quélet
Cap 2.5–5cm/1–2in. The distinctive feature is the tiny white 'teeth' which hang down at the edge of the grey cap. The gills are narrow, greyish-black, mottled with tiny black spots. The stem is slender, grey.
Spores blackish, 13–16 x 8–11µm. On horse or cow dung, widely distributed. Inedible.

Panaeolus semiovatus
(Sow. ex Fr.) Lund. & Nannf.
Cap 2.5–8cm/1–3in. The whitish, egg-shaped cap often cracks in dry weather. The gills are widely spaced, blackish-grey and mottled. The tall stem has a distinct ring or ring-zone at the top. **Spores** blackish, 15–20 x 8–11µm. On horse dung, throughout Britain. Inedible.

Panaeolus foenisecii
(Pers. ex Fr.) R. Maire.
Cap 1–2.5cm/½–1in. The rounded cap starts date-brown but dries out pale buff. The gills are widely spaced, dark brown. The stem is slender, and pale buff. **Spores** dark purple-brown, 12–15 x 6.5–9µm, with minute warts. Very common, scattered in lawns and fields throughout Britain whenever

Psathyrella candolleana
(Fr.) Maire

Cap 2.5–10cm/1–4in. This very delicate mushroom breaks with the least handling, the brittle caps and stems are buffy-brown then almost ivory-white when dry. The gills are narrow, greyish-lavender. **Spores** purplish-brown, 7–10 x 4–5µm. Singly to small tufts often on buried wood in grass or by stumps. Widely distributed in Britain. Edible.

Psathyrella velutina
(Pers. ex Fr.) Singer
= *Lacrymaria lacrymabunda*

Cap 2.5–8cm/1–3in. The reddish-brown cap and stem are both hairy-fibrous and there is a hairy ring-zone at the stem apex. The broad gills are deep yellow to blackish-brown with white edges, and often weeping droplets. **Spores** blackish-brown, 9–12 x 6–7µm. In woods and gardens, usually in disturbed soil along tracks and paths. Edible but best avoided.

Psathyrella hydrophila
(Bull. ex Merat) R. Maire

Cap 5–8cm/2–3in. The reddish-brown cap fades rapidly as it dries from the centre outwards and is very fragile. Tiny fragments of white veil cling to the cap margin. The gills are narrow, crowded, pale brown to chestnut when mature. The slender stem is white, smooth, brittle. **Spores** blackish-brown, 5–7 x 3–4µm. Often found in large tufts on dead timber, common throughout Britain.

Coprinus picaceus
(Bull.) Fries
Cap 5–10cm/2–4in high.
The cap is ovate-cylindrical
when young then expands
into a bell shape when
mature, deep brown at first it
is covered with a fleecy white
veil which soon ruptures to
form numerous patches. The
gills are very narrow,
crowded, white at first then
soon black from the margin
inwards and liquifying to
release the spores. The tall
stem is white, smooth.
Spores black, 14–19 x
10–13μm. An uncommon
species found in beech
woods, especially in southern
England. Edible.

Coprinus comatus Shaggy Mane
(Mull. ex Fr.) S.F. Gray
Cap 2.5–8cm/1–3in.
across, 8–15cm/3–6in.
high. The cap is
unmistakable, tall,
cylindrical and shaggy
white scales, all on a tall,
straight stem with a ring
low down. The entire cap
dissolves away in a few
hours. **Spores** black,
11–15 x 6–8.5μm.
Frequent in disturbed soil
in gardens and along road
edges, throughout Britain.
Edible and good.

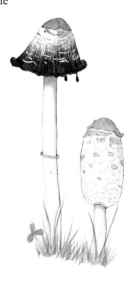

Coprinus atramentarius
(Bull. ex Fr.) Fr
Cap 2.5–8cm/1–3in. The egg-shaped cap is smooth grey-brown with some slight scales at the centre. The white, crowded gills mature black. The white stem has a ridge or ring-zone at the very base. **Spores** black, 7–11 x 4–6μm. As with all *Coprinus* it dissolves away when mature. Very common in clusters near stumps and buried wood in grass, throughout Britain. Edible but reacts with alcohol to produce nausea, flushing and tingling.

Coprinus micaceus
(Bull. ex Fr.) Fr.
Cap 2.5–5cm/1–2in. The whole cap sparkles with minute, mica-like specks of veil when fresh, although they wash off with age. **Spores** black, 7–10 x 4–5μm. Often in very large clusters on or around stumps, throughout Britain. Edible but worthless.

Coprinus plicatilis
(Curtis ex Fr.) Fr.
Cap 0.6–2.5cm/¼–1in. The deeply furrowed cap soon flattens and has a darker, brown centre. The flesh is so thin you can almost see through the cap. The stem is extremely fragile and the mushroom only lasts a few hours. **Spores** black, 10–13 x 6.5–10μm. In grass throughout Britain. Edible.

RUST-BROWN, YELLOW-BROWN, CIGAR-BROWN SPORES

PAXILLUS FAMILY

(Paxillaceae)

This group, despite their gills, are thought to be closely related to the boletes. They share many microscopic and chemical features and have similar spore colours.

Paxillus involutus
(Bat. ex Fr.) Fr.

Cap 5–15cm/2–6in. An important species because it can be poisonous and even fatal in some cases. The distinguishing features are the yellow-brown cap with a woolly, at first inrolled, cap margin, and the soft yellow-brown gills which descend the stem and bruise deep reddish-brown. **Spores** brown, 7–9 x 4–6µm Widespread throughout Britain, very common.

Paxillus atrotomentosus
(Bat. ex Fr.) Fr.

Cap 5–20cm/2–8in. This often very large mushroom is found on stumps of old pine trees and is easily recognized by the brownish cap with decurrent yellow gills, and thick, deep brown, velvety-hairy stem. **Spores** yellowish, 5–6 x 3–4µm. It is common over much of Britain. It is supposedly edible but is best avoided.

Phylloporus rhodoxanthus
(Schw.) Bres.

Cap 2.5–10cm/1–4in. This species has a velvety, reddish-yellow to reddish-brown cap, deep yellow gills which are often cross-veined and wrinkled and a reddish stem. **Spores** brown, 9–12 x 3–5µm. A rare species, mostly in southern England. It is edible but poor in quality.

CORTINARIUS FAMILY

(Cortinariaceae)

This is the largest group of mushrooms in the world and they grow in a bewildering number of forms, colours and sizes. The spores vary from rust brown to dull, cigar-brown and may be smooth, warty or even angular-lumpy. A cobwebby veil is often present.

Cortinarius trivialis
Lange

Cap 5–10cm/2–4in. The sticky cap is yellow-brown while the gills are pale lilac before turning rust-brown. The stem is sticky with rings of sticky, whitish to yellow veil. **Spores** rust-brown, 10–15 x 7–8μm, warty. Frequent in boggy places in deciduous woods, throughout Britain. Inedible.

Cortinarius muscigenus
Peck = *C. collinitus* of some authors

Cap 5–10cm/2–4in. The sticky cap is rich orange-brown to ochre, smooth. The gills are white at first while the stem is flushed with pale blue-violet, and banded with thicker rings. **Spores** rust-brown, 12–15 x 7–8μm, warty. In boggy woods, quite common. Inedible.

Cortinarius pseudosalor
Lange

Cap 5–10cm/2–4in. The very slimy cap is pale ochre-brown and wrinkled while the gills are pale clay-buff. The slimy stem is cylindric, pale bluish-lilac. **Spores** rust-brown, 12–15 x 6–8μm, warty. Very common in beech woods throughout Britain. Inedible.

Cortinarius delibutus
Fries

Cap 5–10cm/2–4in. Often very sticky the cap is bright golden-yellow and smooth. The gills and stem apex are pale lavender when young, the lower stem has a sticky yellow covering. **Spores** rust-brown, 7–8 x 5–6µm, warty. Quite common in boggy woodlands, throughout Britain. Inedible.

Cortinarius croceocaeruleus
(Pers.) Fries

Cap 2.5–5cm/1–2in. The pale blue-lavender to violet cap is sticky when wet and soon fades to ochre at the centre. The gills are ochre. The tapered stem is white to pale ochre, slightly sticky with a faint ring-zone. The taste is distinctly bitter and the odour is sweet, rather like honey. **Spores** rust brown, 7.5–8.5 x 4.5–5.5µm, warty. Fairly common late in the year under beech trees in southern England on chalky soils. Inedible.

Cortinarius volvatus
A. H. Smith

Cap 5–10cm/2–4in. An unusual species because the base of the stem has a thick white veil almost in the form of a volval sac. All parts of the mushroom are bluish when young but the cap turns brownish with age. The taste is bitter. **Spores** rust-brown, 8–10.5 x 5–6µm, warty. A rare species found under beech on chalky soils in the west of England. Inedible.

Cortinarius sodagnitus
Henry

Cap 8–10cm/3–4in. This beautiful species is one of a number that are all violet. The cap fades when old to ochre at the centre and the cuticle is bitter to taste (the flesh is mild). The gills are deep violet when young then rust brown as the spores mature. The stem has a marginate bulb at the base and a ring zone at the top. The flesh is whitish-lavender KOH (caustic soda) placed on the cap turns brilliant red, very distinctive. **Spores** rust-brown, 10–12 x 5.5–6.5μm, warty. Frequent late in the year under beech trees in southern England. Inedible. *C. dibaphus* differs in both flesh and cuticle being bitter and the KOH reaction being pink.

Cortinarius auroturbinatus
(Secr.) Lange

Cap 5–10cm/2–4in. The brilliant golden-yellow to orange cap flushes reddish-orange with age to rust at the centre. The gills are bright yellow then rust-brown. The stem has a marginate bulb and is whitish-yellow with a prominent cobwebby veil when young. The bulb is yellow on the outside. The cap cuticle with KOH turns red-brown to purplish. **Spores** rust-brown, 1.5–15 x 7–8.5μm, lemon shaped, very warty. Quite frequent in beech woods on chalky soils in southern England late in the year. Inedible.

Cortinarius purpurascens
Fries

Cap 5–10cm/2–4in. The dark brown cap is streaked with darker fibres and is sometimes violet at the margin. The gills and stem are violet and both bruise deep violet-purple. The stem has a distinct marginate bulb. **Spores** rust-brown, 8–10 x 4–5.5μm warty. Frequent in mixed woods, throughout Britain. Inedible.

Cortinarius torvus
(Fr.) Fries
Cap 5–10cm/2–4in. A rather dull, brown species distinguished by the whitish veil on the lower stem which looks like a stocking half pulled up. The gills and stem apex are violet when young. **Spores** rust-brown, 8–10 x 5–6µm, warty. Common in beech and oak woods throughout Britain. Inedible.

Cortinarius armillatus
(Fr.) Fr.
Cap 5–10cm/2–4in. The reddish-brown, dry caps are smooth to fibrous. The gills are broad, distant and pale cinnamon-brown. The stem is often very bulbous and has 2–4 narrow bands of bright reddish veil. The spores are rust-brown, 9–12 x 5.5–7.5µm, warty. In mixed woods, especially birch, quite common throughout Britain. Inedible.

Cortinarius evernius
(Fr.) Fr.
Cap 5–10cm/2–4in. The dark, reddish-brown, fibrous cap contrasts with the beautiful violet stem which is banded with white veil and is often long and spindle-shaped. **Spores** rust-brown, 9–11 x 5–6µm, warty. Frequent in wet, or boggy pine woods especially in Scotland. Inedible.

Cortinarius alboviolaceus
(Pers. ex Fr.) Fries
Cap 5–12cm/2–5in. The
specific name means whitish-
violet which is exactly what this
fungus is. The whole
mushroom is silvery white with
a flush of violet or lavender.
The bulbous stem is ringed or
booted with whitish veil at the
base. **Spores** rust-brown, 8–10
x 5–6μm, warty. Common in
mixed woods, throughout
Britain. Inedible.

Cortinarius argentatus
(Pers. ex Fr.) Fries
Cap 5–10cm/2–4in. The
whole mushroom is silvery
white to slightly lilac,
becoming yellowish at the
centre. The stem is often very
bulbous and is almost smooth
and without veil. The gills are
pale brown when mature.
Spores rust-brown, 8–10 x
5–6μm, warty. Not
uncommon in some years in
beech woods in southern
England. Inedible.

Cortinarius bolaris
(Pers. ex Fr.) Fries
Cap 5–8cm/2–3in. The
yellowish cap becomes spotted
with copper-red scales as does
the stem also. The gills are
pale buff becoming rust-
brown. When cut or bruised
all parts turn dull yellowish
then eventually reddish.
Spores rust-brown, 6–7 x
5–6μm, warty. Quite common
in beech woods widely
distributed. Possibly poisonous.

Cortinarius violaceus
(Fr.) S. F. Gray
Cap 5–10cm/2–4in.
Unmistakable by its deep,
blackish-violet fruit-body with
dry, minutely scaly cap and
club-shaped stem. The violet
gills are rust-brown when
mature. **Spores** rust-brown,
13–17 x 8–10µm, warty. Rare,
in beech and birch woods,
widely distributed. Inedible.

Cortinarius rubellus
Cooke = *C. speciosissimus*
Cap 2.5–8cm/1–3in. The cap
shape varies but is usually
bluntly conical and the entire
mushroom is a warm orange-
brown to reddish-orange. The
stem is usually rather spindle-
shaped, rooting and the gills are
widely spaced and broad.
Spores rust-brown, 8–11 x
6.5–8.5µm, warty. A rare
species in conifer woods, often
in moss, mostly in the north.
Deadly poisonous, causes
severe kidney damage.

Cortinarius traganus
(Fries) Fries
Cap 5–10cm/2–4in. The
clear lavender colours, rather
stout fruit-bodies with
smooth cap, and stem with
some whitish veil contrast
with the cut flesh which is
yellow-brown and often
marbled. The odour is
strong and penetrating of
over-ripe pears. **Spores** rust-
brown, 7–10 x 5–6µm,
warty. Frequent in conifer
woods in Scotland. Inedible.

Cortinarius sanguineus
(Wulf.) Fries
Cap 2.5–5cm/1–2in. The
entire mushroom is bright
blood-red to carmine-red,
including the flesh. **Spores**
rust-brown, 6–9 x 4–6μm,
warty. Quite common in
beech woods throughout
Britain. Possibly poisonous.

Cortinarius semisanguineus
(Fr.) Gill.
Cap 2.5–5cm/1–2in. The
yellow-brown, silky cap and
stem contrast with the vivid
cinnabar- to blood-red gills.
The flesh is pale yellow to
orange in the base. **Spores**
rust-brown, 6–8 x 4–5μm,
warty. Common in birch
and pine woods throughout
Britain. Inedible, possibly
poisonous.

Hebeloma crustuliniforme
(Bull. ex St. Amans) Quélet
Cap 5–10cm/2–4in. The
entire mushroom is the
colour of unbaked pastry, a
dull ivory-buff. The gills,
when mature, turn a pale
grey-brown to tan and
often have tiny beads of
moisture on the edges.
There is an odour or radish
and the taste is slightly
bitter. **Spores** dull
cinnamon-brown, 9–13 x
5.5–7μm, minutely warty.
Common in mixed woods
throughout Britain.
Poisonous.

Hebeloma edurum
Metrod ex Bon

Cap 5–15cm/2–6in. The dull, ochraceous cap is smooth with a slightly furrowed margin. The gills are pale buff and do not weep droplets. The stem is rather stout, club-shaped and fibrous-scaly with a slight ring-zone, browning at the base. The odour is fruity or like chocolate, then unpleasant. **Spores** dull cinnamon-brown, 9–12 x 5–7μm, almost smooth. In groups in conifer woods, rather uncommon. Poisonous.

Hebeloma radicosum
(Bull.) Ricken

Cap 8–12.5cm/3–5in. The dull, clay-brown cap has faint, broad, flattened scales and is dry to very viscid when wet. The gills are sinuate, pale ochre. The stem is tapered, deeply rooting in the soil and has a scaly remnants of veil below a distinct ring. The mushroom has a strong odour or marzipan or bitter almonds. **Spores** pale brown, 9–10 x 5–6μm, minutely warted. The fungus arises from the underground nests of small rodents and has even been found originating from a kingfisher's nest. Sadly a rather uncommon species. Probably poisonous.

Rozites caperata
(Pers.) Karst.

Cap 5–12.5cm/2–5in. The pale ochre cap is finely radially wrinkled and dusted with a fine white frosting of veil at the centre. The gills are pale buff and join the stem. The stem is cylindrical, fibrous and has a white ring around the centre and some white fragments of veil at the base. **Spores** rust-brown, 11–14 x 7–9μm, roughened. Frequent in conifer woods Scotland, very rare in England. Edible.

Inocybe calamistrata
(Fr.) Gill.
Cap 2.5–5cm/1–2in. This
dark brown species has a
scaly cap and stem and the
latter has a deep blue-green
flush over the lower half.
The odour is unpleasant,
spermatic–fruity. **Spores**
dull, earthy-brown, 9–12 x
4.5–6.5μm, elliptical.
Common in conifer woods,
often along stream edges,
widely distributed in Britain.
Poisonous.

Inocybe bongardii
(Weinm.) Quélet
Cap 2.5–8cm/1–3in. Pale
cinnamon to pinkish-buff with
darker, flattened scales. The gills
are crowded, pale brown. The
stem is silky, fibrous, whitish-buff
bruising red as does the cap also.
The flesh has a strong odour of
over-ripe pears. **Spores** dull
brown, 10–12 x 6–7μm, bean-
shaped. Frequent in mixed
woods. Poisonous.

Inocybe patouillardii
Bresadola
Cap 5–10cm/2–4in. The white,
bluntly conical cap is fibrous, often
splitting at the margin and stains
bright red when scratched. The gills
are whitish then olive-yellow. The
sturdy stem is white, fibrous, also
bruising reddish. The odour is fruity
to unpleasant. **Spores** dull brown,
10–13 x 5.5–7μm, bean-shaped.
Uncommon, in open grassy
woodlands on calcareous soils, mostly
in southern England. Deadly
poisonous.

Inocybe pyriodora
(Pers. ex Fr.) Quélet

Cap 5–8cm/2–3in. The domed cap is densely fibrous and yellow-brown to pinkish-brown where stained. The gills are whitish to pale brown, crowded, and the stem is silky-fibrous to shaggy. The mushroom has a strong odour of ripe pears . **Spores** 7–10 x 5–7.5µm, elliptical. In deciduous woods, fairly common. Poisonous.

Inocybe geophylla
(Bull.) Karst.

Cap 1–2.5cm/½–1in. The domed cap is silky or smooth and varies from white to pale lilac (var. *lilacina*) as does the stem also. The gills are pale brown and the mushroom has a strong, earthy or spermatic odour. **Spores** dull brown, 7–9 x 4–5.5µm, elliptical. Common in mixed woods throughout Britain. Poisonous.

Gymnopilus junonius
(Fr.) Orton

Cap 8–15cm/3–6in. The bright golden-yellow to tawny-orange cap is coarsely fibrous to slightly scaly. The gills are crowded, shallow, orange to rust-brown and often speckled. The stem is large, tough and fibrous, often very swollen or club-shaped and has a well-developed membranous ring. The taste is extremely bitter. The flesh is yellowish and unchanging. **Spores** bright orange-brown, 8–10 x 5–6µm, coarsely warty. Very common, grows in large clumps at the base of dead deciduous trees. Inedible.

Kuehneromyces mutabilis
(Schff.) Sing. & Smith = *Pholiota mutabilis = Galerina mutabilis*
Cap 2.5–8cm/1–3in. The smooth cap is usually bicoloured with a paler centre as it dries out. The stem is scaly below the distinct ascendant ring. **Spores** rust-brown, 6–7 x 3–4.5μm, smooth. Common, late in the year on stumps of deciduous trees. Edible but must not be confused with the poisonous *Galerina* species.

Galerina marginata
(Fr.) Kuehner
Cap 2.5–5cm/1–2in. The flattened, sticky cap is dark brown to tawny when dry. The gills are attached, crowded and yellowish-rust. The stem is smooth, minutely lined and has a delicate ring above. **Spores** rust-brown, 8.5–10.5 x 5–6.5μm, roughened. In clumps on fallen logs, quite common. Deadly poisonous.

Phaeocollybia lugubris
(Fr.) Heim
Cap 5–8cm/2–3in. The bluntly conical cap is reddish-brown to olive-brown, smooth, with irregular lobed margin. The gills are pale orange-ochre. The stem is smooth, deeply rooting, whitish-brown. **Spores** brown, 7–9 x 4–5μm, warty. A rare species found under conifers, especially spruce in the north. Edibility uncertain.

STROPHARIA FAMILY

(Strophariaceae)

(BROWN-SPORED SECTION)

Most species grow on or near wood and may have glutinous, dry or scaly caps. The gills are sinuate-adnate and the spores are yellow-brown to rust and smooth. Usually a veil or ring is present.

Pholiota lenta

(Pers.) Singer

Cap 5–10cm/2–4in. The pallid, beige cap is very viscid with small white veil fragments near the edge. The gills are pale cinnamon to rust-brown and the stem is club-shaped with woolly scales. The odour is slight, of straw and the taste mild. **Spores** ochre-brown, 6–7 x 3–4μm. In leaf litter and woody debris under beech trees, quite common. Inedible.

Pholiota squarrosa

(Mull.) Kummer

Cap 5–15cm/2–6in. The dry, tawny-yellow cap is covered with recurved, pointed scales. The crowded gills are yellow to slightly olive, then rust-brown. The dry stem is also scaly up to the ring. **Spores** ochre-rust, 6–8 x 3–4μm. Common, in large clumps at the base of deciduous trees. Inedible, possibly poisonous.

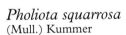

Pholiota flammans
(Fr.) Kummer
Cap 2.5–8cm/1–3in. The brilliant golden-yellow cap is both scaly and sticky. The gills are bright yellow and crowded. The stem is densely scaly up to the ring-zone and is dry. **Spores** ochre-brown, 4–5 x 2.5–3µm. Single to small clusters on conifer logs and stumps, widespread in Britain. Edibility doubtful.

Pholiota aurivella
(Fr.) Kummer
Cap 5–15cm/2–6in. The deep orange-yellow cap is glutinous when wet and has darker spot-like scales which can wash off. The gills are crowded, pale yellow then tawny. The stem is dry, yellowish, with recurved scales. **Spores** ochre-brown, 7–10 x 4.5–6µm. High up on standing trees, or on fallen logs, usually deciduous, widespread. Edibility doubtful.

Pholiota highlandensis
(Peck) Smith & Hesler
Cap 2.5–5cm/1–2in. The sticky cap is a dull reddish-buff to tawny-brown, without scales. The gills are yellowish then rust, slightly decurrent. The stem is tawny and slightly fibrous up to the whitish-yellow ring-zone. **Spores** ochre-brown, 5–8 x 3.5–4.5µm. Always on burn-sites in woodlands, around burned stumps or logs, widely distributed. Edibility doubtful.

Pholiota alnicola

(Fr.) Singer

Cap 5–10cm/2–4in. Very
smooth for this genus, all parts
of the mushroom are yellow to
tawny, with only faint veil
remnants at the cap margin.
There is a fragrant odour,
difficult to define. **Spores**
ochre-brown, 7.5–11 x
4.5–5.5μm. In clumps around
dead trees, especially birch and
alder, frequent. Inedible.

Pholiota gummosa

(Lasch.) Singer

Cap 2.5–8cm/1–3in. The sticky cap is
a pale straw yellow with small whitish
scales at the margin. The gills are pale
straw then ochre-brown. The stem is
coloured like the cap and has woolly,
whitish scales up to a distinct ring-
zone. **Spores** rust-brown, 6–8 x
4–5μm. In small clumps often on clay
soils on buried wood, common.
Inedible.

BOLBITIUS FAMILY

(Bolbitiaceae)

These fungi have their cap cuticle made up of rounded, instead of thread-like cells, and all have pale brown spores.

Bolbitius vitellinus
(Pers.ex Fr.) Fr.

Cap 2–5cm/¾–2in. The bright lemon-yellow cap is sticky or slimy and deeply grooved when old, and very fragile. The gills are bright cinnamon-brown and often liquefy in wet weather. The pale yellow-white stem is minutely hairy. **Spores** rust-ochre, 10–15 x 6–9µm, smooth. Found on dung, compost or old straw, common throughout Britain. Edible but too fragile.

Agrocybe praecox
(Pers. ex Fr.)

Cap 2.5–8cm/1–3in. The tan-brown cap becomes paler when dry, is smooth and sometimes cracks in dry weather. The gills are attached and pale buffy-brown. The stem is slender and has a thin, membranous ring above. **Spores** dark brown, 8–11 x 5–6µm. Found in woods, flower beds in woodchip mulching, gardens, in spring and early summer, widely distributed. Inedible.

Agrocybe molesta
(Lasch) Singer = *A. dura*

Cap 5–10cm/2–4in. The whitish to pale tan cap soon cracks with age and dry weather. The gills are attached, crowded and pale buff then dark brown. The rather stout stem is stiff, smooth, whitish with a delicate ring which soon vanishes. **Spores** dark brown, 10–14 x 6.5–8µm, smooth. Often in groups, in grass or along roadsides and waste places. Common in early summer throughout Britain. Edible.

Conocybe tenera

(Schaeff. ex Fr.)
Cap 1–2.5cm/⅜–1in. Conical to bell-
shaped, the cap is rich yellow-brown
to reddish-brown, fading when dry.
The thin gills are almost free, widely
spaced and cinnamon-brown. The
tall, thin stem is pale brown and very
finely lined top to bottom. **Spores**
reddish-brown, 11–12 x 6–7μm,
smooth. In lawns, grassy edges of
woodlands, common, throughout
Britain. Inedible.

PINK SPORES

The pink-spored mushrooms have spores which range
from pale pink to deep salmon, almost brownish-pink.

TRICHOLOMA FAMILY

(Tricholomataceae)

(PINK-SPORED SECTION)

Although the majority of species in this family are white-
spored, there are some exceptions, of which this is an
example. Because the spores have a distinct pinkish hue,
it is placed here in the guide, with other pink-spored
genera, where one might expect to look first.

Rhodotus palmatus

(Bull. ex Fr.) Maire
Cap 2.5–8cm/1–3in. The beautiful
reddish-pink to apricot cap has a
thick, rubbery-gelatinous cuticle
which is strangely wrinkled and
pitted. The gills are attached to
the stem and pale pink. The stout
stem is pink and fibrous, rather
tough, and usually set off-centre.
Spores pink, 6–8μm, globose and
warty. On dead deciduous stumps
and logs, widespread but uncommon. Inedible.

ENTOLOMA FAMILY

(Entolomataceae)

The *Entoloma* mushrooms and their relatives all have strange, angular pink spores, or long spores with angular ridges.

Clitopilus prunulus
(Scop. ex Fr.) Kumm.
Cap 2.5–8cm/1–3in. The white to pale grey cap has a texture like kid-leather, and is often irregular and wavy in outline. The decurrent gills are white the pink. The short stem is smooth, white, often off-centre. There is a strong odour and taste of fresh ground meal, bread dough or cucumber. **Spores** pink, 10–12 x 5–7μm, with longitudinal ridges, appearing angular in end view. On soil in woods, throughout Britain. Edible.

Entoloma serrulatum
(Pers.) Hesler
Cap 2.5–5cm/1–2in. The blue-black to almost black cap is dry and fibrous, and is depressed at the centre. The gills are pale blue with darker, almost black edges. The stem is coloured like the cap or paler, smooth. The flesh is almost odourless. **Spores** pink, 7–10 x 5–7.5μm, angular, many-sided. Quite common in grass and woody debris. Inedible.

Entoloma incanum

(Fr.) Hesler

Cap 2–2.5cm/²/₃–1in. An unusual fungus because of its clear green coloration. The cap is olive-green, depressed at the centre and finely striate. The gills are white to pale green. The stem is slender, smooth and a bright blue-green bruising darker. The odour is strong and unusual, of mice or burnt hair. **Spores** pink, 11–14 x 8–9µm, angular, 7–12-sided. Quite common in open pastures on calcareous soils. Inedible.

Entoloma sericeum

(Bull.) Quélet

Cap 2.5–8cm/1–3in. The fleshy, umbonate cap is dark reddish-brown to grey-brown, silky-shiny and with the margin noticeably striate. As the fungus dries out the cap becomes much paler. The gills are whitish-grey then pink. The stem is coloured like the cap and is fibrous with a white base. The flesh has a strong mealy taste and smell. **Spores** pink, 7.5–10 x 6.5–8µm, angular. A common species in grassland and lawns everywhere. Inedible.

Entoloma porphyrophaeum

(Fr.) Karsten

Cap 5–8cm/2–3in. A tall, fleshy mushroom it has a domed cap of a soft, grey-violet to pinkish-lilac in colour. The cap surface is finely fibrous and felty. The gills are broad, greyish then pink. The tall stem is coloured like the cap and is fibrillose with a white base. **Spores** pink, 10–13 x 6–7µm, angular. In grassy clearings and meadows, quite common in the south. Inedible.

Entoloma clypeatum
(L.) Kummer

Cap 5–10cm/2–4in. The dull, grey-brown to ochre-brown cap is dry, smooth to slightly fibrous or streaky, often cracked and irregularly split at the margin. The broad gills are sinuate, pink. The robust stem is white, tough, fibrous and the flesh has the taste and odour of meal or flour. **Spores** pink, 8–11 x 7.5–9μm, angular and many-sided. Quite common in the spring and early summer in grass under roses, hawthorn bushes. Inedible.

Entoloma lividum
Quélet

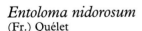

Cap 5–15cm/2–6in. A large, fleshy species, the cap is ivory-white to pale buff, smooth, often slightly streaky or mottled. The gills are sinuate, yellow-ochre when young then pink when mature. The stout stem is fibrous, whitish. The odour and taste are strong of fresh meal or flour. **Spores** pink, 8–10 x 7–9μm, 5–7-sided. In grassy woodlands on calcareous soils, widespread. Poisonous, causes severe gastric upsets, can be deadly.

Entoloma nidorosum
(Fr.) Quélet

Cap 5–8cm/2–3in. The slightly umbonate or even depressed cap is pale ochre to greyish-ochre, smooth and very hygrophanous, becoming much paler as it dries. The gills are whitish-grey then pink. The tall stem is whitish, striate and the flesh has a strong nitrous or bleach-like odour when fresh. **Spores** pink, 7–9 x 6–8μm, angular. Common in wet woodlands, especially near willows, throughout Britain. Probably poisonous.

PLUTEUS FAMILY

(Pluteaceae)

These all have gills completely free of the stem and deep
salmon-pink spores. One genus (*Volvariella*) has a volva
at the base of the stem.

Pluteus cervinus
(Schaeff. ex Fr.) Kummer
Cap 5–10cm/2–4in. The dull grey-brown
to dark brown cap is smooth but with fine
fibres and occasionally some tiny scales at
the centre. The gills are quite free of the
stem and start white, then mature pink.
The fibrous stem is white above shading
to brown below. **Spores** pink, 5.5–7 x
4–5μm, smooth. This very common
mushroom grows on dead wood, sawdust
piles and buried wood throughout Britain.
Edible but tasteless. On the edges of the
gills are bottle-shaped cystidia with 2–3
hook-like projections.

Pluteus petasatus
(Fr.) Gillet
Cap 5–15cm/2–6in. This almost white species
has a faint flush of brown in its cap, which is
fibrous-scaly at the centre. The gills are often
very deep and rounded at the margin, and usually
stay pale for a long time before turning pink when
mature. The stout stem is fibrous, slightly brown
below. **Spores** pink, 6–10 x 4–6μm. Common on
old deciduous stumps and on sawdust or
woodchips, widely distributed. Edible and good.

Pluteus salicinus
(Britz.) Sacc.
Cap 2.5–5cm/1–2in. The deep brown, rather
wrinkled or veined cap contrasts with the clear
yellow of the stem. The young gills are also pale
yellow before turning pink. **Spores** pink, 6–7 x
5–6μm, smooth. Quite a common species on old
logs of deciduous trees, widespread throughout
Britain. Edibility uncertain.

Pluteus aurantiorugosus
(Trog.) Sacc.
Cap 2.5–5cm/1–2in. The brilliant
scarlet cap fades to orange as it
expands, while the gills when young
are yellow. The stem is whitish above
but orange-yellow below. **Spores**
pink, 5.5–6.5 x 4–4.5μm. This small,
but spectacular species occurs on
dead elm and maple but is sadly not
common. Edible.

Volvariella bombycina
(Schaeff. ex Fr.) Singer
Cap 5–15cm/2–6in. This
magnificent mushroom has a silky-
shaggy cap of white tinged with
yellow. The broad, free gills are
pink and the white stem emerges
from a large, egg-like volva.
Spores pink, 6.5–10.5 x
4.5–6.5μm. It grows out of holes,
or wounds in the trunks of elms,
maples and beech trees and
sometimes on old stumps. It is a
widespread but rather uncommon
species. Edible and good.

Volvariella speciosa
(Pers. ex Fr.) Singer
Cap 5–10cm/2–4in. The smooth,
white to greyish cap is often very
sticky or viscid. The gills are free,
broad, and pink when mature. The
often long stem is white and comes
from a thick, deep volval cup. **Spores**
pink, 11.5–21 x 7–12μm. Quite
common in some years in fields in old
grass, rotting straw or around stables.
Edible, but because it grows on the
ground, and has a volva, one might
easily pick a poisonous *Amanita*
instead. It is safer to buy the canned
Paddy-straw Mushrooms *(V.
volvacea)*.

GASTEROMYCETES

PHALLALES, STINKHORNS AND SQUID FUNGI
PUFFBALLS AND EARTHBALLS
BIRDS NEST FUNGI

Although this is now considered by most mycologists to be a rather artificial grouping of fungi which are not really closely related, it is nevertheless a very convenient one for use in guide books, since it brings together all those fungi which produce their spores inside the fruit-body, rather than on an external hymenium.

The Gasteromycetes rely to a great extent on external forces – wind, rain, insects – to carry off their spores and have evolved some of the strangest forms to accomplish this.

PHALLALES, STINKHORNS AND SQUID FUNGI

All species emerge from an 'egg' and expand rapidly in a matter of hours to full size. Their spore-mass (gleba) liquifies and gives off an unpleasant odour to attract insects which eat the spores and unknowingly pass them on to germinate elsewhere. All the stinkhorns can be 'hatched' by placing large, unopen eggs on a damp paper towel under a glass.

Phallus hadriani
Vent.
10–15cm/4–6in high. The egg is tinted pinkish-mauve and the spongy stem is white. The thimble-like cap is pitted and ridged with a large disc-like opening at the apex. Covering the cap is the dark green spore mass which liquifies and has an odour of rotting flesh or vegetables. **Spores** 4–5 x 2μm. An uncommon species found in Britain only in sand-dunes by the sea and other similar maritime habitats.

Phallus impudicus Stinkhorn
Linnaeus

10–15cm/4–6in high. Starting out as a
large white egg it soon emerges to
form tall spongy stem with a thimble-
like cap on top. The cap is pitted and
ridged and covered in a dark olive
spore-mass. The apex of the cap has a
circular opening. The stem is white,
hollow. Inside the egg is a pale amber
jelly which surrounded the stem and
cap as they developed. In a very short
time the spore-mass liquifies and
releases a foul odour which attracts
insects who both eat the spores and
disperse them accidentally. **Spores**
4–5 x 2μm. Very common in woods
and even gardens everywhere. Edible
when young.

Mutinus caninus
(Huds.) Fries

8–12.5cm/3–5in high. This
slender, delicate species
emerges from a small white
egg to produce a white
spongy stem with a bright
red or orange apex which is
covered in a dark olive spore
mass. There is no separate
cap present as in the
previous species. The spores
liquify and give off a very
faint odour, sometimes
almost undetectable. **Spores**
3–4 x 1.5μm. Quite
common, usually in clusters
in beech woods. Inedible.

Clathrus ruber Cage or Lattice Stinkhorn
Mich. ex Pers.
5–16cm/2–6in high. Once again, an
unmistakable fungus with its thick,
orange-red, cage-like structure.
Spores on the inner surface of the
cage and are strongly fetid. The egg is
white, often with mycelial cords at the
base. **Spores** 5–6 x 1.5–2.5μm. It is
found in a few scattered localities in
gardens and shrubberies in the south
of England and in the Isle of Wight.
Probably introduced but now
naturalised. Inedible.

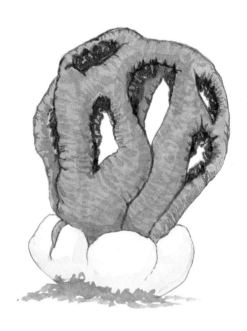

PUFFBALLS AND EARTHBALLS

As their common name suggests, they are ball-like and the spores which are inside the ball become dry and powdery, and will puff out if the ball is tapped.

Rhizopogon luteolus
Fries

2.5–5cm/1–2in. Forming a round to oval fruit-body it consists of a yellow-ochre to olive-yellow skin forming a tough outer wall. There are numerous thin, brown mycelial strands adhering to the outer surface. The spore-mass (gleba) inside is a dull olive-brown. **Spores** 7–10 x 2.5–3.5µm. It grows in sandy pine woods often just pushing up through the soil, and is quite common. Inedible.

Vascellum pratense
(Pers.) Kreisel

2.5–5cm/1–2in across. The small whitish to yellowish-buff ball has a scurfy, powdery covering of tiny spines which easily rub off. The base of the ball forms a short, sterile stem. The spore mass starts white but soon matures to a dark olive-brown. **Spores** 3–5.5µm, globose, finely warted. Common in fields, lawns and golf-courses everywhere. Edible when young.

Lycoperdon echinatum
Pers. ex Fries

2.5–5cm/1–2in across. The pale brown puffball is remarkable for the coating of long, pointed spines. These are in groups with their tips touching and can be almost 5mm (¼in) long. These will rub off to leave a dark network pattern on the cuticle. The spore mass matures to a dark purple-brown and emerges through an opening at the top of the ball. **Spores** brown, 4–6μm, warted. Rather uncommon in deciduous woods throughout Britain. Edible when young.

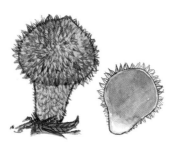

Lycoperdon mammiforme
Persoon

5–8cm/2–3in across. The white to pale ochre ball has a short stalk and is unusual for the way in which the outer wall breaks up to form numerous cottony scales and patches leaving a ring-like zone around the base. The spore mass is dark purple-brown when mature. **Spores** brown, 4–5μm, globose, warted. A rare species of woodlands on chalky soils. Edibility unknown.

Lycoperdon pyriforme
Schaeff.· ex Pers.

2–4.5cm/⅝–1¾in across. These are rather pear-shaped puffballs and start white before soon turning yellowish-buff. The skin is rather smooth. At the bottom of the very short sterile base are white strands running into the rotten wood on which it grows. **Spores** olive-brown, 3–4.5μm, globose and smooth. Often grows in large clusters on dead wood, common and widely distributed. Edible when young and white.

Lycoperdon perlatum
Pers.

2.5–5cm/1–2in across. The white, rounded to slightly club-shaped ball has small white spines or warts, often in tiny rings which are easily rubbed off leaving round marks. A short to rather distinct sterile base is present. The spore-mass is white then greenish-ochre when mature. **Spores** 3.5–4.5μm, globose and minutely warted. Common in fields, roadsides and gardens, widely distributed. Edible when young and if white and completely uniform in appearance.

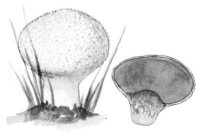

Lycoperdon foetidum
Bonord.

2.5–5cm/1–2in across. This is the only puffball to have tiny, dark brown spines, even when young. The whole puffball matures to a dark yellowish hue, contrasting with the blackish spines. The spore-mass is dull sepia-brown. **Spores** 4–5µm, globose and minutely spiny. Common in woods, especially conifers, widely distributed. Edible when young, it smells rather unpleasant when old, as do many other species.

Calvatia excipuliformis
(Pers.) Perdek

8–15cm/3–6in high. This large puffball forms a ball on a distinct stem which can elongate to form a tall stalk. The colour changes from white when young to dull brown when old. The outer surface is covered in a fine, granular, warty layer which rubs off. The spore mass is dark olive-brown when mature. **Spores** olive, 3.5–5.5µm, warted. Common in woods everywhere. Edible when young.

Calvatia utriformis

(Bull. ex Pers.) Jaap

8–12.5cm/3–5in across. This large, squat puffball has hardly any stem, the white to pale brown skin has a layer of cottony, scurfy warts which soon break up and fall off. The spore mass is olive-brown when mature and the spores are released by the slow break-up of the entire outer skin. **Spores** olive, 4–5µm, warted. Common in fields and heathland especially in the north, when the ball has broken up the short sterile base remains behind for many months afterwards. Edible when young.

Bovista plumbea

Pers. ex Fries

2–3cm/²⁄₃–1in across. Forming a simple greyish-white ball which turns duller grey with age. The outer surface flakes off in large scales to expose a lead coloured inner layer which encloses the olive-brown spores. **Spores** olive, 4.5–6µm, globose with a very long pedicel or stalk. Quite common in short grass everywhere. Edible when young.

Langermannia gigantea Giant Puffball
(Batsch.) Rotsk. = *Calvatia gigantea*
20–50cm/8–20in across. This often huge species (specimens as large as a small sheep have been found) forms a smooth white ball or flattened oval with a skin the texture of kid-leather. As it matures this flakes away to expose the yellow-brown spore-mass. **Spores** 3.5–5.5μm, globose, minutely warted. Quite common in some years in hedgerows, fields and gardens, widespread. Edible and good when young.

Geastrum triplex Earthstar
Jung.
2.5–8cm/1–3in across. The leathery, grey-brown, onion-like ball splits open to form a star with a pinkish-buff, smooth to cracked surface. At the centre is a ball containing the spore-mass, this is set in a distinct, cup-like collar. **Spores** dark brown, 3.5–4.5μm, globose, warted. In leaf-litter in woods, especially beech, widely distributed. Inedible.

Geastrum pectinatum
Persoon

2.5–5cm/1–2in across. The grey-brown, tough ball splits to form a star, at the centre is a rounded grey-white ball set on a slender stalk. The ball as a narrow, beaked, deeply grooved opening on the top where the spores are released. **Spores** brown, 4–6μm, globose, warted. On the ground in woods and gardens, especially near cedars, widespread but uncommon. Inedible.

Scleroderma citrinum Common Earthball
Persoon

5–10cm/2–4in across. The earthballs differ from the true puffballs in their thick, leathery skins and spores which do not all mature simultaneously. This species has a yellow-ochre, warty-scaly thick skin and a spore-mass which starts white but soon becomes marbled with purple-black. The skin flakes away slowly over many weeks to release the spores. **Spores** 8–12μm, with spines and a fine network. The mature fungus has a strong, pungent odour difficult to define. It grows on soil in woodlands throughout much of Britain. Poisonous.

Scleroderma verrucosum
(Bull.) Persoon

2.5–8cm/1–3in across. The tough earthball is raised up on a thick, rooting stem. The outer skin of the ball is a pale ochre-brown with small, darker scales. The leathery wall is not as thick as in the previous species and breaks open to expose the olive-brown spore mass inside, **Spores** 10–14μm, globose and with tiny sharp spines. Common in sandy woods throughout Britain. Poisonous.

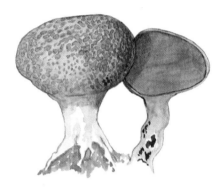

Astraeus hygrometricus
(Pers.) Morgan

2.5–5cm/1–2in across. Starting as a dark brown, onion-like ball in the soil it splits to form a star, exposing a rounded, grey-white ball at the centre. The spore-mass in the ball is brown. **Spores** 7–11μm, warted. The arms of the 'star' become cracked as they bend back and will close up if the weather turns dry, only to reopen when it becomes moist again. Rare, in sandy soils, often sand-dunes, starting out buried then pushing up through the soil. Inedible.

Pisolithus tinctorius Dye–makers Mushroom
(Pers.) Coker & Couch

5–10cm/2–4in across, 5–20cm/2–8in high. A rather ugly fungus, it is often buried deep in the soil. It forms an irregular club-shaped structure, ochre to reddish-brown with a variable length of stem. The spore-mass forms a mass of small, whitish to yellow eggs embedded in a blackish jelly, eventually forming a brownish powder. This crumbles away with time leaving the sterile base in the soil. **Spores** 7–12μm, spiny. Very rare, only known from a few sandy pine woods in southern England although difficult to spot and possibly under-recorded. Inedible. The fungus is boiled down to make a rich golden-brown to black dye.

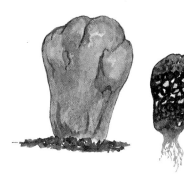

Tulostoma brumale
Persoon

Head about 2cm/⅝in across, stem about 4cm/1½in high. Like a small puffball on a slender, bulbous stem, this species grows in dry, sandy soils. **Spores** brown, 3–5μm, globose, minutely warted. Locally common. Inedible. There are a number of other very similar species, differing in minor details.

BIRDS NEST FUNGI

As the common name suggests, they look like tiny birds'
nests with minute eggs inside. The eggs are in fact the
spore masses.

Cyathus striatus Birds-Nest Fungus
(Huds.) Willd.
0.6–2cm/¼–⅝in across. The urn-shaped structure is hairy on
the outside and with a smooth, grey, fluted interior. At the
bottom are tiny dark eggs attached by coiled cords. These eggs
are the spore masses and are thrown out of the cup by
splashing raindrops. **Spores** 15–20 x 8–12μm, smooth but
notched at one end. Quite common in woodlands on fallen
twigs and branches.

Cyathus olla Common Birds-Nest
(Batsch.) Pers.
0.6–1cm/¼–⅜in across. This funnel-shaped species is brown
and roughened on the outside, but smooth and white on the
inside. The eggs are greyish. **Spores** 11–13 x 7–8μm, smooth.
Common in fields, gardens, attached to woody debris, widely
distributed.

Crucibulum laeve

(Huds.) Kamb.
0.6–1cm/¼–⅜in across. The almost
cylindrical nest is tawny-yellow with a
hairy lid covering it at first. This splits
open to reveal the smooth, inner
surface. The eggs are whitish. **Spores**
4–10 x 4–6μm, smooth. On dead
wood and woody debris, widely
distributed.

Sphaerobolus stellatus

Tode
1.5–2.5mm across. This tiny fungus is
included here even though not really
qualifying as one of the 'larger fungi'
because of its remarkable system of
spore dispersal. The tiny whitish eggs
split open to form a yellowish star
with a brownish ball of spores at the
centre. The inner layer of the star flips
outward suddenly to propel the spore-
mass into the air up to a distance of
4m (13ft). This is a common (if rarely
observed) species on sawdust, dung
and other organic debris.

CHANTERELLES

(Cantharellaceae)

These often edible fungi include the well-known Chanterelle eaten all over the world. They all lack true gills, forming their spores on either a smooth undersurface or on blunt wrinkles or ridges.

Craterellus cornucopiodes

(L. ex Fr.) Persoon
2.5–8cm/1–3in across. These black, funnel-shaped fungi are very thin and completely hollow down the centre. The inner surface may be slightly scaly and becomes pale grey when fry. The outer surface is a whitish-grey when mature. **Spores** white, 10–11 x 6–7μm. Common in damp deciduous woodlands everywhere, usually on mossy banks in large numbers. Edible and good.

Pseudocraterellus sinuosus

(Fr.) Corner
2.5–5cm/1–2in. The wavy, irregular caps are often split and are coloured a pale grey-brown. They form a hollow funnel-shape with the outer surface smooth to slightly veined and greyish-pink. **Spores** ochre, 10 x 8μm. An uncommon species found on damp, mossy banks and leaf litter in mixed woods.

Cantharellus cibarius Chanterelle
Fries

2.5–15cm/1–6in across. The yellow-orange cap is often wavy and irregular in shape, with an inrolled margin, and is very fleshy. The undersurface is formed of numerous ridges and wrinkles, often cross-veined and descending the short, stout stem. The flesh is white. The odour is pleasant of apricots. **Spores** pale buff, 8–11 x 4–6μm, smooth. In groups in woods of oak, birch or pine, common throughout Britain. Edible and delicious.

Cantharellus tubaeformis
Fries

2.5–8cm/1–3in across. The yellow-brown cap is thin and often depressed at the centre. The blunt, wrinkles and ridges on the underside are yellow then greyish-violet and run down the stem. The stem is slender, hollow and yellow-orange. **Spores** cream, 8–12 x 6–10μm, smooth. Common in sphagnum moss in boggy woods in throughout Britain. Edible and good.

Cantharellus cinereus

2.5–8cm/1–3in. This thin fleshed, grey-black mushroom is funnel-shaped with the undersurface distinctly wrinkled-veined running down the stem. It might be mistaken for the Black Trumpet *C. cornucopiodes* but that species is smooth on the outside. **Spores** white, 8–10 x 5–7μm. An uncommon to rare species, found in damp, mossy woods. Edible.

Cantharellus lutescens

(Pers.) Fries

2.5–5cm/1–2in. The slender fruit-bodies form a hollow trumpet with the inner surface a yellow-orange to yellow-brown, rather scaly at centre. The outer surface is a brighter yellow-orange, with shallow, veins and wrinkles running down onto the yellow stem. **Spores** white, 10 x 7μm. A rare species restricted almost entirely the Scottish highlands under pine. Edible.

CORAL AND CLUB FUNGI

As their common name suggests, these fungi resemble undersea corals or coloured clubs; many are among the brightest coloured of fungi and can reach large sizes. Their spores are produced on basidia on the outer surface of each club.

Ramaria aurea
(Fr.) Quélet

5–15cm/2–6in across. This coral-like fungus is a bright golden yellow-orange with masses of closely packed branches. Each branch ends in a cluster of tiny cauliflower-like tips. The central stem is paler, almost white. **Spores** deep ochre, 8–15 x 3–6µm, roughened. A rather rare but spectacular species it grows in beech woods in the south. Edible.

Ramaria formosa
(Fr.) Quélet

10–20cm/4–8in across. This large coral mushroom varies from a pinkish-orange to coral-pink but always has yellow tips to the numerous branches. The branches arise from a large central stem. The flesh is white to pale orange and often bruises wine to purplish-black. **Spores** ochre, 8–15 x 4–6µm, roughened. A rare species it grows in leaf-litter in deciduous woods especially beech. Not edible, causes gastric upsets and diarrhoea.

Ramaria ochraceovirens

(Jungh.) Donk = *R. abietina*
5–10cm/2–4in. This species has
numerous tangled branches
arising from a common central
stem. The branches may be
flattened to rounded and are
coloured ochre to yellow-olive
when young. As the fungus ages
the entire fruit-body stains
olive-green, also when handled.
Spores ochre, 9–10.5 x
3.5–5μm, roughened. A
common species in the litter of
pine woods. Inedible.

Ramaria stricta

(Fr.) Quélet
2.5–10cm/1–4in across. This yellow-
brown coral has many straight,
parallel branches held very upright.
The stems fuse below into a central
stalk, the flesh is brownish-white. All
parts bruise darker on handling. The
odour may be unpleasant and the
taste bitter. **Spores** golden-yellow,
7–10 x 3.5–5.5um, minutely warted.
Widespread on coniferous logs, quite
common. Inedible.

Ramaria botrytis

(Fr.) Rick.
5–15cm/2–6in across. A beautiful,
cauliflower-like coral with many
densely branched clubs, whitish
with pinkish-purple tips. The
flesh is white. **Spores** pale
ochre, 11–17 x 4–6um, with
longitudinal lines. Under
beech, widely distributed.
Edible but easily confused
with other less edible species,
i.e. *R. formosa* has pinkish
branches with yellowish tips.

Pterula multifida
(Fries) Fries

1-3in/2.5-7.5cm high. This coral fungus consists of clusters of vertical, densely crowded and very fine branches up to 1mm thick. There is a common central stem at the base from which they arise. In colour they vary from pale cream to ochre brown becoming lilac-brown when old. Unlike many corals they are rather tough and elastic in texture. **Spores** are white, 5-6 x 2.5-3.5μm, smooth. On needle litter or conifer wood in conifer forests, not common but widespread. Inedible.

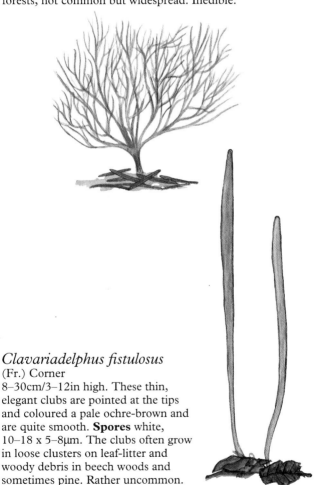

Clavariadelphus fistulosus
(Fr.) Corner

8–30cm/3–12in high. These thin, elegant clubs are pointed at the tips and coloured a pale ochre-brown and are quite smooth. **Spores** white, 10–18 x 5–8μm. The clubs often grow in loose clusters on leaf-litter and woody debris in beech woods and sometimes pine. Rather uncommon. Edible.

Clavariadelphus pistillaris
(Fr.) Donk
8–20cm/3–8in high,
2.5–5cm/1–2in across.
Forming a single, large
club, the colour ranges
from pale ochre to
pinkish-brown and it
bruises darker brown.
The flesh is white and
bitter to taste. **Spores**
creamy-white, 11–16 x
6–10μm, smooth. In leaf-
litter in beech woods,
widespread in southern
England. Inedible.

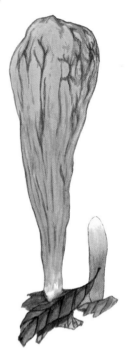

Clavariadelphus truncatus
(Quel.) Donk
5–15cm/2–6in high, 2.5–8cm/1–3in
across. Forming a flat-topped,
wrinkled club from pinkish-brown
to yellow-ochre in colour. The flesh
is white and mild to the taste.
Spores ochre, 9–13 x 5–8μm,
smooth. Under conifers, a rare find
but very easy to recognize with its
flattened cap. Inedible.

Sparassis crispa Cauliflower Mushroom

Wulf. ex Fr.

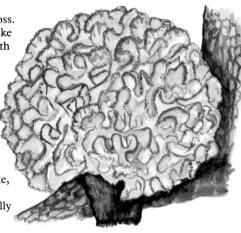

15–30cm/6–12in across. Looking very much like a large cauliflower with numerous flattened, twisted and crisped, leaf-like branches, pale cream-yellow in colour. The odour is slightly spicy. It often has a rooting base buried deep in the ground. **Spores** white, 5–7 x 3–5μm. At the base of conifers, locally common and widely distributed. A good edible species.

Clavulinopsis fusiformis

(Sow.) Corner

5–15cm/2–6in high. This golden-yellow fungus forms a clump of simple, pointed clubs, hollow inside. **Spores** creamy, 5–9 x 4.5–9μm, smooth. Common in grassland and open woods, widely distributed. Edible.

Clavaria vermicularis

Fries

5–15cm/2–6in high. The Latin name means worm-like and this does suggest a cluster of white, wavy worms. The clubs are hollow and very brittle. **Spores** white, 5–7 x 3–4μm, smooth. In grassy places and open woodlands, widely distributed. Edible. The similar species *C. fumosa* differs in its greyish-lilac clubs.

Clavulina cristata
(Fr.) Schroet.
2.5–8cm/1–3in across and high. A densely branched, coral-like species with a central stem and finely pointed tips it varies from pure white to yellowish. **Spores** white, 7–11 x 6.5–10µm, nearly round and smooth. In mixed woodlands throughout Britain Edible. The similar *C. cinerea* has blunter, greyish-white branches.

Clavulina amethystina
(Fr.) Donk
2.5–10cm/1–4in across and high. A beautiful lilac-violet coral-like species with many branched clubs. **Spores** white, 7–12 x 6–8µm, smooth and growing on 2-spored basidia. In deciduous woods, rare. Edible.

Thelephora terrestris
Fr.
2.5–8cm/1–3in across. Forming a cluster of overlapping, fan-shaped or vase-shaped caps, dark, blackish-brown and hairy-fibrous. The margin in often whitish and torn. The underside is smooth to wrinkled or minutely lumpy. There is usually an earthy, mouldy odour. **Spores** purplish-brown, 8–12 x 6–9µm, minutely spiny. Common on soil in mixed woods, widely distributed. Inedible.

CRUST FUNGI

These form crust-like sheets on wood and occasionally spread out into small shelves or brackets. They do not have tubes on the undersurface like the true bracket fungi (Polypores).

Chondrostereum purpureum Silver Leaf Fungus
(Fr.) Pouz.
10–15cm/4–6in across or larger. This fungus forms irregular, leathery sheets with fused, overlapping caps and is coloured a lovely bright lavender-purple on the edge and undersurface, a dull yellow-ochre above. **Spores** white, 5–6.5 x 2–3µm, smooth. On apples, plum and other fruit and deciduous forest trees, common. This fungus attacks its host, causing it to show a characteristic silvery blight on the leaves, and finally to sicken and die.

Stereum hirsutum
(Willd. ex Fr.) S.F. Gray
0.6–2.5cm/¼–1in across. Many of the flattened, shell-like caps may be fused together in rows, each one is concentrically zoned in shades of brown, cream or grey and the upper surface is minutely hairy. The lower surface is smooth, yellow-grey. **Spores** white, 5–8 x 2–3.5µm, smooth. On dead deciduous wood, especially birch and beech, widely distributed. Inedible.

TOOTHED FUNGI

These fungi all form their spores on projecting conical teeth which usually hang down from the undersurface of the cap or fruit-body. Some species grow on the ground and have a stem, others form shelf-like brackets on trees.

Hydnellum spongiosipes
(Pk.) Pouz.

Cap 5–10cm/2–4in. The very tough, almost woody cap is convex to flattened and rich reddish-brown, minutely hairy. The tiny, densely crowded spines on the undersurface are dark brown and run down the short, very bulbous, velvety stem. Several caps may be fused together. The flesh is two-layered with a dark, zoned, inner core. **Spores** brown, 5.5–7 x 5–6µm, warted. Uncommon, in groups in leaf-litter under oaks, widely distributed. Inedible. There are numerous other species, many with strong tastes and odours and coloured flesh, i.e. *H. caeruleum* with flesh banded with blue and orange-brown stem flesh, and mealy odour.

Sarcodon imbricatum
(L.) Karst.

Cap 10–15cm/4–6in. The grey-brown cap is coarsely scaly and the undersurface has densely packed, tiny grey-brown teeth which run down the short stem. The flesh is soft, pale brown and may taste a little sharp. **Spores** brown, 6–8 x 5–7μm, with large warts. Under conifers in sandy soils, rather rare, mostly in Scotland. Edible. Other species include *S. scabrosum* with blackish-green stem base and very bitter taste, and *S. joeides* with pale lavender flesh which turns green with KOH solution.

Hydnum repandum
Linnaeus

Cap 5–10cm/2–4in. The smooth, irregularly shaped cap is a pale orange-buff, the teeth are white to pale pinkish-yellow and descend the short, whitish-yellow stem. The flesh is white bruising orange-brown and has a mild to slightly bitter taste. **Spores** white, 6.5–9 x 6.5–8μm, nearly round, smooth. Common, often in groups in deciduous woods, especially beech, sometimes under conifers, throughout Britain. Edible.

Hydnum rufescens
Fries

Cap 1–2.5cm/½–1in. A small, reddish-brown species with the spines less decurrent than in *H. repandum* and with a well-formed, slender, and central stem. **Spores** 8–10 x 6–7µm. Uncommon, found in pine woods and sometimes deciduous, often considered to be a variety of the previous species by some authors. Edible.

Hericium ramosum
(Bull. ex Mer.) Let.

10–25cm/4–10in across. This beautiful, large, but delicate species is like an undersea coral growing on a tree. Composed of a mass of narrow arms which repeatedly branch and have short spines all along the lower surfaces. **Spores** 3–5 x 3–4µm, white, finely roughened. Frequent on old deciduous logs throughout Britain. Edible and delicious.

Hericium erinaceus
(Fr.) Pers.
10–25cm/4–10in across. Forming a compact oval ball, all the
spines are in a dense mass and are quite long (about 2.5–5cm/
1–2in) **Spores** white, 5–6.5 x 4–5.5µm, minutely roughened.
This attractive white fungus grows, often very high up, on
living trees of beech, oak and maple, widely distributed.
Edible and delicious when young.

Auriscalpium vulgare Earpick Fungus
S.F. Gray
Cap 1–2.5cm/³⁄₈–1in. The
cap of this small species is
dark brown, minutely hairy
and with the slender, hairy
stem attached to one edge.
The tiny spines are whitish-
brown. **Spores** white, 5–6 x
4–5µm, minutely spiny.
Found only on fallen, rotting
pine-cones, this small, dark
species is often overlooked.
Widespread throughout
Britain. Inedible. The generic
name is Latin for earpick, an
instrument used by the
Romans for personal hygiene.

POLYPORE FUNGI, BRACKET AND SHELF FUNGI

All fleshy or woody fungi with their spores produced inside tubes (except for the boletes) are included here. Many are very large and woody and are common on our woodland trees. Some grow on the ground and can look like a bolete but are much tougher fleshed. A number are serious parasites of trees.

Polyporus brumalis Winter Polypore
Fries

Cap 2.5–10cm/1–4in. One of a number of polypores with a cap and more or less central stem, this species has a smooth, yellow-brown cap, whitish tubes with pores spaced 2–3 per mm, and a slender grey-brown, minutely hairy stem. **Spores** white, 5–7 x 2–2.5µm, sausage-shaped, smooth. On fallen timber, especially birch from late autumn throughout the winter and into spring, widespread. Inedible.

Polyporus varius

Fries = *P. elegans*

Cap 5–10cm/2–4in. The cap is pale cinnamon-buff, smooth, and the tubes and pores are yellowish to pale brown, 4–5 per mm. The almost central stem is slender, pale tan with a black lower half. **Spores** white, 7–10 x 2–3.5μm, sausage-shaped, smooth. On fallen twigs and logs of deciduous wood throughout Britain. Inedible. *P. badius* has a larger, reddish-brown, shiny cap and often completely black stem.

Polyporus squamosus

Fries

Cap 10–30cm/4–12in. A magnificent species looking rather like a seat or saddle growing out of dead or standing trees. The cinnamon-brown cap has broad, darker brown, flattened scales, and the pale cream-buff pores are large, honeycomb-like and descend the short, tough stem. The latter has a black base. **Spores** white, 10–16 x 4–6μm, cylindrical, smooth. Widespread in Britain. Edible and good when very young.

Grifola frondosa Hen of the Woods
(Fr.) S.F. Gray

Individual caps 2.5–8cm/1–3in,
entire fruit-body 15–50cm/6–20in.
This large fungus has numerous
semicircular caps fused together into
one large structure with a central
stem. The caps are grey to brown,
finely woolly-fibrous, fleshy,
with the pores white, small
and angular, descending
the stem. The flesh is
solid and white. **Spores**
white, 5–7 x 3.5–5µm,
smooth. At the base of
old oaks and other
deciduous trees, often
where the tree has
been struck by lightning.
Common in southern
England. Edible and
delicious.

Meripilus giganteus
(Fr.) Kar.

25–75cm/10–30in across. This huge
fungus has many broad shelves fused
into a single mass. The shelves are
fleshy, rather soft, yellow-ochre to
tan, smooth to finely velvety with
blunt margins.
The pores
are white.
All parts
bruise black.
Spores white,
6–7 x 4.5–6µm,
smooth.
Growing at
the base of dead or
living deciduous trees,
especially beech and oak,
quite common throughout
Britain. Edible.

Laetiporus sulphureus Chicken Mushroom

(Fr) Murr.
Shelves 10–75cm/4–30in across.
Growing in large, overlapping
masses, each shelf is soft and
fleshy, bright orange-yellow to
salmon above and with lemon-
yellow pores. **Spores** white,
5–7 x 3.5–5µm, smooth.
Growing on dead or dying trees,
both deciduous and conifers it
can form huge clusters weighing
many pounds, often high up in the
tree but also at the base. Edible and
delicious and quite unmistakable. There
is a variety with pinkish cap and white pores,
semialbinus, which is rather uncommon.

Postia caesia Blue-cheese Polypore

(Schr.) Karst. = *Tyromyces caesius*
Shelf 5–15cm/2–6in across. This
rounded, rather lumpy species is soft
and fleshy and starts greyish-white
but soon flushes pale blue when
rubbed or bruised. The pores
are small, white and also
bruise bluish. The odour is
fragrant and the taste mild.
Spores pale blue, 4–5 x
0.7–1µm, sausage-like and
smooth. Common on dead
wood throughout Britain.

Hapalopilus nidulans

(Fr.) Kar.
Cap 2.5–12.5cm/1–5in. The deep
orange-brown fruit-body is fleshy,
thick, with small, cinnamon-brown
pores. The flesh is thick and watery,
tawny-brown. All parts turn bright
purple-violet with KOH solution.
Spores white, 3.5–5 x 2–3µm,
smooth. Common on fallen logs of
deciduous trees, throughout Britain.
Inedible.

Piptoporus betulinus Birch Polypore
(Fr.) Kar.

Cap 5–25cm/2–10in. The
rounded, kidney-shaped caps
are white to pale tan,
smooth, with an inrolled
margin. A short stem is often
present joining the bracket to
the tree. The pores are
extremely small and white.
Spores white, 5–6 x 1.5µm,
sausage-shaped. Found only
on birch trees it is a serious
parasite. Inedible. The soft,
white flesh has a number of
uses: fire lighting, sharpening
razors, halting bleeding, and
today is cut in narrow strips
to pin insects in displays.

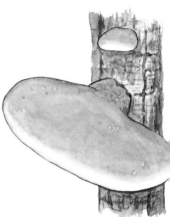

Trametes versicolor
(Fr.) Pilat

Cap 2.5–10cm/1–4in. Usually in
overlapping clusters, each bracket is
thin, semicircular and with
multicoloured zones, from brown to
grey, blue-black, yellow to green. The
fine pores are pale yellowish. **Spores**
white, 5–6 x 1.5–2.2µm, sausage-
shaped, smooth. Very common on
fallen timber throughout Britain.
Inedible.

Daedaliopsis confragosa
(Fr.) Schroer.
Cap 5–10cm/2–6in. The
semicircular to kidney-
shaped brackets are tough,
broad at the back with a thin
margin, grey-brown and
zoned. The pores vary from
round to quite maze-like or
in extreme forms even gill-
like; they are pale cream-buff
bruising pinkish. **Spores**
white, 7–11 x 2–3μm,
sausage-shaped, smooth.
Very common on fallen
timber, often persisting for
several years, throughout
Britain. Inedible.

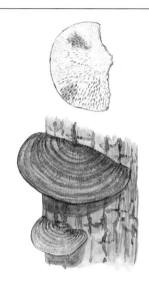

Daedalea quercina
Fries
Cap 5–15cm/2–6in. The thick, very
tough bracket is smooth to rather
cracked or furrowed on top, white to
pale ochre, often slightly zonate. The
pores are very elongated, maze-like
with thick walls, whitish-buff. **Spores**
white, 5.5–7 x 2.5–3.5μm, cylindrical,
smooth. Frequent on dead oaks,
throughout Britain. Inedible.

Ganoderma applanatum Artist's Fungus
(Pers. ex Wall.) Pat.

Cap 10–50cm/4–20in. The very hard, woody brackets are thick, tapering to the blunt margin, lasting for many years and adding tube layer to tube layer each year. The upper surface is smooth to lumpy, crust-like, grey-brown with a powdery covering of brown spores (which are attracted and held by a static charge formed on the surface). The margin is white and the pores on the underside are cream bruising instantly brown when scratched. The flesh is pale cinnamon with small whitish flecks. **Spores** reddish-brown, 6.5–9.5 x 5–7µm, with a thick double wall, perforated on the outer layer. Uncommon, on deciduous trees both living and dead, throughout Britain. Scratching the pores with a toothpick, you can draw a picture which will become permanent as the fungus dries.

The much more common *G. adspersum* has darker brown flesh without white flecks and larger spores 8–13µm long.

Ganoderma lucidum
(Leyss.) Karsten
10–25cm/4–10in. The beautiful glossy brackets vary from deep purplish-red to brown or purple-black, while the actively growing margin can be a bright whitish-yellow. The surface of fresh specimens look as if lacquered with varnish. The pore surface is whitish-buff bruising darker. The bracket may have a distinct and sometimes long stem. **Spores** brown, 7–13 x 6–8µm, ovate with a flattened end. This is a fairly common fungus found on oak, chestnut and other deciduous trees especially in the south. Inedible.

Fistulina hepatica Beefsteak Polypore
Schaeff. ex Fr.

Cap 8–25cm/3–10in. The tongue-like to semicircular bracket is soft and fleshy, gelatinous and often slimy on top, usually minutely roughened, pimpled, deep blood-red. The tubes are reddish with white pore mouths and are easily separable from each other. The flesh is fibrous, wet and exudes a reddish-brown liquid. **Spores** pinkish-salmon, 4.5–6 x 3–4μm, smooth. Quite common, found on oaks and chestnut, widespread. Edible but rather acidic, an acquired taste. The mushroom often drips 'blood' and looks very liver or tongue-like indeed.

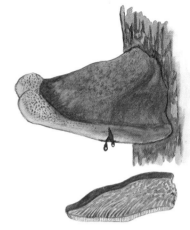

Fomes fomentarius
(L. ex Fr.) Kickx

5–20cm/2–8in. The thick hoof-shaped brackets are hard and woody, with a concentrically grooved crust. The colour varies from grey-brown to pale brown or even blackish. The pores are dull brownish, bruising darker. The flesh is hard, cinnamon-brown with an acrid taste. **Spores** yellowish, 15–20 x 5–7μm. Commonest in Scotland and Northern England where it grows on birch, but also found in the south mostly on beech and sycamore. Inedible.

Schizophyllum commune
Fries

Cap 1–5cm/⅜–2in. The small, thin caps are white and hairy. The undersurface looks at first like gills but are gill-like folds split lengthwise on the edges, pale pinkish-buff. **Spores** white, 3–4 x 1–2.5μm, smooth. On fallen branches of deciduous trees, widespread. Found throughout the world this species can be found all year, surviving droughts by rolling up the folds until wet weather returns. Inedible.

JELLY FUNGI AND RELATED GROUPS

All these fungi belong to a large group which share the common feature of having their spore-producing cells, the basidia, divided (septate) either longitudinally or transversely. They are usually of a jelly-like or rubbery consistency.

Tremella mesenterica Witches' Butter
Ret. ex Fr.
2.5–10cm/1–4in across. A common species throughout Britain and throughout the year whenever the weather is damp, it grows on twigs and fallen branches. It forms an irregular, lobed, jelly-like mass but when the weather turns dry it shrinks to a hard, horny orange lump, reviving when it rains again.
Spores yellowish, 7–15 x 6–10μm. Basidia are divided longitudinally.

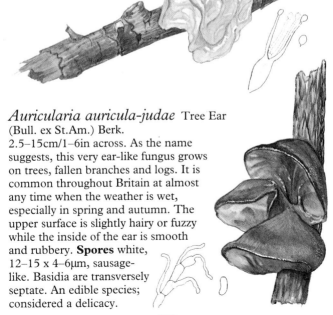

Auricularia auricula-judae Tree Ear
(Bull. ex St.Am.) Berk.
2.5–15cm/1–6in across. As the name suggests, this very ear-like fungus grows on trees, fallen branches and logs. It is common throughout Britain at almost any time when the weather is wet, especially in spring and autumn. The upper surface is slightly hairy or fuzzy while the inside of the ear is smooth and rubbery. **Spores** white, 12–15 x 4–6μm, sausage-like. Basidia are transversely septate. An edible species; considered a delicacy.

Tremiscus helvelloides

(Pers.) Donk = *Phlogiotis helvelloides*
(Fr.) Mar.
2.5–10cm/1–4in across. The beautiful pink to apricot, often
funnel-shaped, wavy or lobed fruit-bodies, are quite
unmistakable. They grow in clusters on the ground under
conifers or on very rotted wood in the fall. The flesh is
rubbery. **Spores** white, 9–12 x 4–6µm. Basidia
are longitudinally septate. Edible but rather tasteless.
A very rare but distinctive species apparently
increasing its range in Britain in recent years.

Calocera viscosa

(Pers. ex Fr.) Fr.
2.5–10cm/1–4in high. This
coral-like mushroom is bright
golden-yellow and has a tough,
gelatinous texture, unlike most
of the true Coral fungi. The
branches fork at the tips and
may be blunt to quite pointed.
Spores ochre-yellow, 9–14 x
3–5µm, sausage-shaped,
smooth. The basidia are shaped
like a tuning-fork. Common on
dead coniferous wood
throughout Britain. Inedible.
C. cornea differs in its very
small, simple, unforked clubs
which are pale yellow and grows
on deciduous wood.

ASCOMYCETES

CUP FUNGI AND
RELATED FUNGI

These fungi produce their spores in a quite different way from all the other fungi shown in this book. Their spores are contained in a special cell called an ascus, usually 8 spores per cell although this can vary depending on the genus and species concerned. The spores are ejected from the ascus like bullets from a gun, often producing a cloud of spores above the fungus when it is picked or disturbed in any way. The ascus gives its name to the group as a whole – the Ascomycetes.

CUP FUNGI AND RELATED FUNGI

Morchella esculenta Yellow Morel
L. ex Fries

Cap 5–12.5cm/2–5in high. These highly sought after mushrooms are very distinctive with their spongy caps set on swollen stems. The cap begins tightly compressed and a dark blackish-brown with white ridges; as it expands it turns to a bright yellow-ochre and the ridges become thin and jagged. The microscopic asci with their spores line the pits and ridges. The stem is white and very rough or granular, and is usually swollen at the base. The entire mushroom is hollow when cut in half. **Spores** (8 per ascus) are deep yellow-ochre in deposit, 20–24 x 12–24µm, elliptical, smooth. Found in the spring for a few weeks, under dying apple, elm and ash trees. A wonderful edible species often growing in enormous numbers where conditions are right.

Morchella elata Black Morel
Fries

Cap 5–12.5cm/2–5in high.
A rather conical, narrow
species with often very
regular, parallel ribs with
cross ridges, the cap is a dark
smoky, blackish-grey,
sometimes paler in the pits
when mature. The stem is
usually cylindrical, white and
roughened. The cap and
stem are hollow. **Spores**
pale ochre, 24–28 x
12–14μm, elliptical, smooth.
Often connected with
conifers or ash, found only in
the spring, widespread.
Edible and delicious.

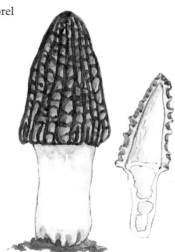

Morchella hortensis
Boudier

Cap 5–10cm/2–4in high. The
dark, grey-brown to blackish
cap has noticeably parallel
ridges, where the cap meets the
stem there is almost no space or
depression as in some other
species and the cap has a
curiously flattened, ovate
appearance. The stem is cream
to pale buff, swollen at the base
and like the cap completely
hollow. **Spores** pale yellow,
19–22 x 15–16μm. This is an
uncommon species and unusual
in preferring urban and other
disturbed habitats rather than
woodlands or orchards. It
frequently appears in flower
pots in greenhouses and
gardens. Edible and delicious.

Morchella conica
(Pers.) Boudier
Cap 5–8cm/2–3in high.
Distinctive by its broad,
conical shape, rather parallel
ridges and no separation
between the cap and stem.
The colour varies from grey-
brown to ochre-brown when
mature. The stem is usually
short and squat. **Spores**
medium yellow, 20–25 x
15–17μm. An uncommon
species in Britain, found in
pine woods and also open
grassland with shrubs on
chalky soils. Edible
and delicious.

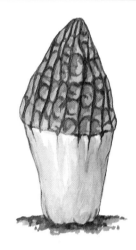

Verpa conica
Swartz ex Pers.
Cap 1–3cm/½–1¼in
high. The thimble-
shaped cap is smooth,
not wrinkled, and only
attached to the long
stem at the very apex.
The smooth to granular
stem is white to yellow-
ochre and completely
hollow. **Spores** pale
ochre, 22–26 x
12–16μm, elliptical,
smooth, and 8 per
ascus. Frequent in
shady woodlands and
old orchards,
widespread. Edible but
not very choice.

Mitrophora semilibera
DC. ex Fries

Cap 2.5–5cm/1–2in high.
The conical to blunt cap is
dark brown to yellow-brown
with darker, almost parallel
ridges and is attached to the
stem for about half of its
height. The stem is small at
first but may soon expand to
become very tall and swollen
in relation to the cap, and is
whitish with a granular
texture. The mushroom is
completely hollow. **Spores**
pale ochre, 24–30 x
12–15µm, elliptical, smooth.
Common and usually the
first morel to appear, under
mixed deciduous trees, early
spring, widespread. Edible
but rather thin-fleshed.

Helvella crispa
Scop. ex Fr.
2.5–8cm/1–3in high. A
beautiful, snow-white
species with a
curiously fluted,
columned stem and
saddle-shaped, thin
cap. The whole fungus
is very fragile and
crisp, the stem is
hollow with several
convoluted chambers.
Spores cream, 18–21
x 10–13µm, elliptical.
Frequent in open,
grassy woodlands,
widespread. Edibility
uncertain.

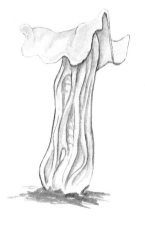

Helvella lacunosa
Afz. ex Fries
5–10cm/2–4in high. The entire fungus is an ashen, sooty grey-black with the stem fluted, twisted and ribbed, and the saddle-like thin cap often lobed and downcurved over the stem. The stem is hollow and chambered. **Spores** pale cream, 15–20 x 9–12µm, elliptical, smooth. Common in mixed woods, widespread. Edibility doubtful.

Gyromitra infula False Morel
(Schaeff. ex Fr.) Quélet
Cap 2.5–10cm/1–4in high. A dark reddish-brown, the saddle-like, thin cap clasps the rounded, cylindrical stem. The latter is flushed pale brown and is hollow with some chambers. **Spores** cream, 18–24 x 7µm, elliptical, smooth. Uncommon, on fallen conifer logs or debris, mainly in Scotland. Poisonous. The very similar *G. ambigua* has violet tints in its brown cap and stem.

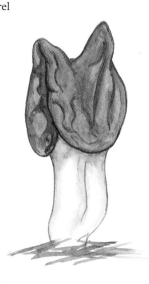

Gyromitra esculenta False Morel
(Pers. ex Fr.) Fr.

Cap 5–10cm/2–4in high. The
rounded, very wrinkled,
almost brain-like cap is a
shiny reddish-brown, the
interior is hollow and
chambered. The stem is
white to pale brown and
smooth to granular. **Spores**
cream, 18–22 x 9–12μm,
elliptical, smooth. In
the spring under conifers,
widespread but
uncommon. Poisonous
and deadly, it contains a cumulative poison,
a chemical similar to that used in rocket fuel.

Otidea onotica
(Pers.) Fuckel

Cup 2.5–5cm/1–2in. The long, rabbit
ear-like cups are a bright apricot
yellow, darker, more orange inside
(which is lined with the microscopic
asci). One or more 'ears' may be
fused together from a common base.
Spores 12–14 x 5–6μm, elliptical,
smooth. On soil and leaf-litter in
mixed woods, uncommon,
widespread. Edible. *O. leporina* is
similar but a dull yellow-ochre.

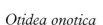

Peziza vesiculosa
Bull.

Cup 2.5–8cm/1–3in. The yellow-
ochre cup is paler, whitish and very
rough, scurfy-granular on the outer
surface. The edges of the cup are
often jagged, tooth-like and
incurved. **Spores** 20–24 x
12–14μm, elliptical, smooth.
Not common but often in
large groups on old
compost, straw, manure
heaps, widespread. Inedible.

Peziza michelii
(Boudier) Dennis
Cup 2.5–10cm/1–4in. Found in spring and early summer this dark, reddish-brown cup fungus is flushed with violet on the outer surface and often olive on the inside. **Spores** 17–21 x 8–10µm, elliptical, finely warted. An uncommon species but very attractive with its violet tints, found in deciduous woods. Edible.

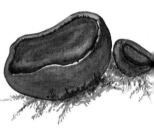

Aleuria aurantia Orange Peel
(Fr.) Fuckel
Cup 2.5–10cm/1–4in. The flattened cups really can be mistaken for pieces of orange peel (and vice versa). The inner surface is smooth and brilliant orange while the outer surface is minutely scurfy-hairy and paler, whitish-orange. **Spores** 18–22 x 9–10µm, elliptical with a network of ridges. Common in clusters on soil by paths, roads and disturbed areas, widespread. Edible.

Scutellinia scutellata
(L. ex Fr.) Lamb.
Cup 0.6–1cm/¼–½in. The deep scarlet cup is fringed with short, black 'eyelashes' all around the margin. The undersurface of the cup is minutely hairy. **Spores** 18–19 x 10–12µm, elliptical, warty. Very common on wet, rotten logs or soil, throughout Britain. This is a complex of closely related species differing mostly microscopically.

Trichoglossum hirsutum Earth Tongue
(Pers.) Boud.

Club 2.5–8cm/1–3in high.
This all-black, flattened club
is minutely hairy all over (use
a hand lens). The club is often
grooved or twisted. **Spores**
enormously long and thin,
100–150 x 6–7µm, brown,
divided into 16 chambers.
Common, often in large
numbers on the forest floor or
in sphagnum moss, widely
distributed. Inedible. There
are many other species
differing in their spores and
the number of chambers
therein. The genus *Geoglossum*
looks identical but is
completely smooth.

Leotia lubrica Jelly Babies
Pers. ex Fr.

Club 2.5–5cm/1–2in high.
Forming a rounded,
gelatinous head on a
slender, rubbery stem. The
cap is ochre-yellow to
olive-yellow while the stem
is yellow-orange with
darker, greenish dots or
granules. **Spores** 17–26 x
4–6µm, spindle-shaped.
Common, in small clusters
in leaf-litter throughout
Britain.

Chlorociboria aeruginascens
(Nyl.) Kan.
Cup 0.6cm/¼in. The thin, flattened cup is a vivid blue-green, almost unique amongst cup fungi. The mature cups are rather rare, it is much commoner to find the wood on which it grows stained throughout its length with bright blue-green as if dyed. **Spores** 6–10 x 1.5–2µm, spindle-shaped. Common on fallen branches, especially oak, throughout Britain.

Bulgaria inquinans Rubber Buttons
(Pers.) Fries
1–5cm/½–2in. The common name of rubber buttons is very descriptive since they look and feel like black rubber buttons. The outer surface may be brownish and roughened while the inner is smooth and black. The asci contain 8 spores, the upper 4 brown and kidney-shaped, the lower 4 colourless but similarly shaped. On fallen branches and trunks of oaks and beeches, common. Inedible.

Xylaria hypoxylon
(L. ex Hook.) Grev.
Club 2.5–8cm/1–3in. high. The thin, flattened, and often branched clubs look like blackened antlers with the base a sooty black but the tips white. The white powder is formed of asexual spores, later it will turn black. **Spores** 11–14 x 5–6µm, bean-shaped, black. Very common on dead wood, throughout Britain.

Xylaria polymorpha Dead Man's Fingers
(Pers. ex Mer.) Grev.
Club 5–10cm/2–4in high. The
swollen, irregular clubs are
usually single but in groups
from a common base; they are
black and roughened, granular.
They are sometimes white and
powdery with asexual spores.
The flesh is thick, white.
Spores 20–30 x 5–10µm,
spindle-shaped, black.
Common on dead wood,
throughout Britain.

Cordyceps ophioglossoides
(Ehr. ex Fr.) Link
Club 5–10cm/2–4in
high. This slender club
has a distinct, reddish-
brown, granular head set
on a yellow, smooth
stem. If the stem is
carefully followed down
into the soil, bright
golden threads will be
found at the base and
these are attached to
another fungus, a false
truffle *Elaphomyces*, a
round, granular reddish-
orange ball. **Spores** of
Cordyceps are 2.5–5 x
2µm, elliptical, smooth.
Uncommon, on the
forest floor of deciduous
woodlands, widespread.

Appendix

Useful Chemicals

Ferrous sulphate ($FeSO_4$) 10% solution in water, very important in a wide range of mushrooms, especially the genus *Russula*. The reaction varies from pink to grey, bluish or dark green.

Ammonia (NH3) 5% solution in water (houshold glass cleaners with ammonia are quite sufficient). Very useful, reactions range from bright red to violet or green.

Potassium Hydroxide (KOH) 10% solution in water, positive reactions are usually yellow to orange or reddish-brown.

Melzer's Solution Formula is: distilled water 20cc, potassium iodine 1g, iodine 0.5g, chloral hydrate 20g. Essential in the study of *Russula* and *Lactarius* and many other fungi if the microscope is to be used.

VERY IMPORTANT! These chemicals are poisonous and/or corrosive. Great care should be taken when handling.

Mushroom societies

Many natural history societies may have a separate mushroom club or group, to find out where to join, consult your local library.

Index

RSNC

The Royal Society for Nature Conservation is pleased to endorse these excellent, fully illustrated pocket guide books which provide invaluable information on the wildlife of Britain and Europe. Royalties from each book sold will go to help the RSNC's network of 48 Wildlife Trusts and over 50 Urban Wildlife Groups, all working to protect rare and endangered wildlife and threatened habitats. The RSNC and the Wildlife Trusts have a combined membership of 184,000 and look after over 1800 nature reserves. If you would like to find out more, please contact RSNC, The Green, Whitham Park, Lincoln LN5 7NR. Telephone 0522 752326.